COMPUTER CONTROL OF MANUFACTURING SYSTEMS

COMPUTER CONTROL OF MANUFACTURING SYSTEMS

Yoram Koren

Technion—Israel Institute of Technology

McGraw-Hill Book Company

New York   St. Louis   San Francisco   Auckland   Bogotá   Hamburg
Johannesburg   London   Madrid   Mexico   Montreal   New Delhi
Panama   Paris   São Paulo   Singapore   Sydney   Tokyo   Toronto

To my wife, Aliza,
my father, Shlomo,
and in loving memory of my mother, Bathia

This book was set in Times Roman by Benj. H. Tyrrel.
The editors were Rodger H. Klas and Susan Hazlett;
the cover was designed by Albert M. Cetta;
the production supervisor was Diane Renda.
The drawings were done by VIP Graphics.
R. R. Donnelley & Sons Company was printer and binder.

COMPUTER CONTROL OF MANUFACTURING SYSTEMS

234567890 DOCDOC 89876543

ISBN 0-07-035341-7

Library of Congress Cataloging in Publication Data

Koren, Yoram.
    Computer control of manufacturing systems.

    Includes bibliographies and index.
    1. Production engineering—Data processing.
2. Machine-tools—Numerical control.   3. Robots,
Industrial.   I. Title.
TS176.K6515   1983   670.42'7   82-14950
ISBN 0-07-035341-7

CONTENTS

PREFACE

The decline in productivity has been one of the main concerns of American industry in the early 1980s. To increase productivity, industry has tried to apply more computerized automation in manufacturing. This has led to an increased number of computer-controlled machine tools, an appearance of industrial robots in the production lines, and the introduction of new technologies such as laser-beam machining.

The revolutionary change in factory production techniques and management that is predicted to take place by the end of the twentieth century will require unprecedented involvement of computer-controlled systems in the production process. Every operation in this factory of the future, from the product design, to manufacturing, assembly, and product inspection, will be monitored and controlled by computers, and performed by robots and intelligent systems. However, this trend toward computerized manufacturing is leading to a demand for appropriately trained engineers to design and maintain these systems. In response to this demand, the industry has established centers for manufacturing, productivity, and robotics at major U.S. universities, such as the University of Michigan, Carnegie-Mellon, RPI, MIT, etc., with the objective of educating more engineers in the fields of modern manufacturing. Nevertheless, to educate students one needs appropriate textbooks as the basis for the development of a curriculum in the required areas. Such books are not currently available for modern manufacturing. In addition to supporting the above-mentioned centers, U.S. industry should encourage and support the writing of textbooks in the field of modern manufacturing.

The purpose of this book is to provide an introduction to the theory and applications of control in the manufacturing area. The book presents concepts of computer control as applied to stand-alone manufacturing systems, such as machine tools and industrial robots, and provides a useful approach to their implementations.

The book introduces the varied aspects of computer control of manufacturing systems:

Machines and mechanical hardware
Part programming languages
Algorithms for interpolation and control
Basics of digital control loops
Adaptive control and optimization
Industrial robots
Flexible manufacturing systems

It is unique in the sense that it covers all these aspects, while other existing books treat only certain limited areas of computerized systems. In addition, the coverage of interpolation, control loops, and adaptive control is more thorough than in other books.

The book has been used as a text for senior undergraduate students in mechanical and industrial engineering at the University of Michigan, and consequently is directed mainly toward this type of audience. The text, however, is written in a self-study format and can be readily used by engineers in industry who wish to adapt their knowledge to the growing importance of computerized systems.

At the University of Michigan the text is used in a single semester course with 2 h of lectures and 3 h of laboratory per week, yielding a total of 28 lectures. The laboratory is divided into two parts: (1) experiments in digital circuits, such as logic gates and counters; (2) programming and manufacturing of parts on a NC lathe and a CNC milling machine. The outline of the course at the University of Michigan is given in column A of the table below. The reading assignment and advanced material (i.e., additional reading for advanced students) in the table are related to this outline.

Chapter No.	Recitation time (hours)			Reading assignment	Advanced material
	A	B	C		
1	2	3	3	1-5	
2	1	2	2	2-5, 2-6	
3	5	7	7	3-3.2, 3-3.5	
4	3	5	3	4-1.1, 4-2.3, 4-3	
5	4	5	2	5-4	5-5
6	4	6	1		6-3.5
7	2	3	1	7-2, 7-3, 7-6	7-5
8	3	3	2	8-6	8-4, 8-5
9	3	3	3		
10	1	2	2	10-4, 10-5, 10-6	
Total	28	39	26		

Other universities might teach the course without the laboratory. For these universities I propose two outlines: column B for 3 h, and column C for 2 h of recitation per week. Since the laboratory is eliminated, more time is required to teach part programming (Chap. 3). The longer course emphasizes the hardware, interpolation, and control aspects of manufacturing systems, whereas the shorter course is more oriented to the traditional manufacturing approach, with a relatively higher percentage of the lectures devoted to machining and programming. Obviously, other approaches could be adopted depending upon the students' motivation and ability.

A course in programming is not a prerequisite for the understanding of any portion of the book. A general background in machining and use of cutting tools is desired. The mathematical knowledge necessary for the understanding of parts of Chaps. 6 and 7 is the Laplace transformation technique which is summarized in the appendix.

A number of examples and homework problems are given for each chapter of the book. A solution manual is available from the publisher for university professors.

ACKNOWLEDGMENTS

The author has had a great deal of help and guidance during his professional career from many colleagues at various universities. The author wishes to thank Professors J. Ben-Uri (Technion, Israel), J. G. Bollinger (Wisconsin, Madison), E. Lenz (Technion, Israel), R. Levi (Politecnico di Torino, Italy), S. Malkin (Technion, Israel), H. Mergler (Case Western Reserve), J. Peklenik (Ljubljana, Yugoslavia), D. T. Pratt (Washington, Seattle), and J. Tlusty (McMaster, Canada) for their encouragement and valuable assistance.

The book was written while the author served as the Goebel Chair Professor of mechanical engineering and Director of Integrated Design and Manufacturing Division at the Center for Robotics and Integrated Manufacturing, University of Michigan. The manuscript has been reviewed by Professors S. Malkin (Chaps. 2 and 8), G. Ulsoy (Chaps. 4 through 6), M. Zarrugh (Chap. 9), and R. C. Wilson (Chap. 10) at the University of Michigan, as well as by two reviewers at the McGraw-Hill Book Company, and the author wishes to thank them all for their thorough review and valuable suggestions.

In acknowledging the help of others in writing the book the author would like to thank his former students O. Masory, M. Shpitalni, G. Amitay, and M. Green for their valuable research work in CNC and adaptive control systems for manufacturing. A major part of the drawings in the book was performed by Mrs. Z. Kalmar, and the manuscript was typed by Mrs. C. Cooper and Ms. L. Hagerman.

Finally, I thank my wife, Aliza, for her encouragement, day after day, and my children, Shlomik and Esther, for their patience through the year it took to complete this book.

Yoram Koren

LIST OF ABBREVIATIONS

ac=alternate current
AC=adaptive control
ACC=adaptive control constraint
ACO=adaptive control optimization
ADC=analog-to-digital converter
ALU=arithmetic and logic unit
APT=automatically programmed tools
ATC=automatic tool changer
BLU=basic length-unit
BRU=basic resolution-unit
BTR=behind the tape reader
CAD=computer-aided design
CAM=computer-aided manufacturing
CIM=computer-integrated manufacturing
CLU=control loops unit
CNC=computer(-ized) numerical control
CPU=central processing unit
CRT=cathode-ray tube
DAC=digital-to-analog converter
dc=direct current
DDA=digital differential analyzer
DPU=data processing unit
DRS=data reduction subsystem
EB=end of block

ECG=electrochemical grinding
ECM=electrochemical machining
EDM=electrical discharge machining
FF=flip-flop
FMS=flexible manufacturing system
FPU=floating-point unit
FRN=feedrate number
IAE=integral of absolute error
I/O=input-output
ITM=improved Tustin method
LSB=least-significant bit
LSI=large-scale integration
MCU=machine control unit
MRR=material removal rate
MSB=most-significant bit
MT=machine tool
NC=numerical control
PM=permanent magnet
pps=pulse per second
PTP=point-to-point
RAM=random-access memory
ROM=read-only memory
r=revolution
rad=radian
rpm=revolution per minute
RWS=rewind-stop
TTL=transistor-transistor logic
TWR=tool wear rate
2-D=two-dimensional

INTRODUCTION

The declining cost of minicomputers and microcomputers is changing the look of the factory floor. Although the application of computers to manufacturing has been somewhat slow, distinct trends can be observed. These include an increase in the use of computer-controlled machine tools, the application of new manufacturing systems such as laser-beam cutters, and the appearance of a new generation of industrial robots in the production lines.

1-1 BASIC CONCEPTS IN MANUFACTURING SYSTEMS

Modern manufacturing systems and industrial robots are advanced automation systems that utilize computers as an integral part of their control. Computers are now a vital part of automation. They control stand-alone manufacturing systems, such as various machine tools, welders, and laser-beam cutters. They run production lines and are beginning to take over control of an entire factory. Even more challenging are the new robots performing various operations in industrial plants and participating in the full automation of factories.

It is well to keep in mind that the automatically controlled factory is nothing more than the latest development in the industrial revolution that began in Europe two centuries ago and progressed through the following stages:

1. Construction of simple production machines and mechanization started in 1770, at the beginning of this revolution.
2. Fixed automatic mechanisms and transfer lines for mass production came along at the turn of this century. The transfer line is an organization of manufacturing

1

facilities for faster output and shorter production time. The cycle of operations is simple and fixed and is designed to produce a certain product.

3. Next came machine tools with simple automatic control, such as plugboard controllers to perform a fixed sequence of operations and copying machines in which a stylus moves on a master copy and simultaneously transmits command signals to servodrives.

4. The introduction of numerical control (NC) in 1952 opened a new era in automation. NC is based on digital computer principles, which was a new technology at that time.

5. The logical extension of NC was computerized numerical control (CNC) for machine tools, in which a minicomputer is included as an integral part of the control cabinet.

6. Industrial robots were developed simultaneously with CNC systems. The first commercial robot was manufactured in 1961, but they did not play a major role in manufacturing until the late 1970s.

7. A fully automatic factory which employs a flexible manufacturing system (FMS) and computer-aided design/computer-aided manufacturing (CAD/CAM) techniques is the next logical extension. FMS means a facility that includes manufacturing cells, each cell containing a robot serving several CNC machines, and an automatic material-handling system interfaced with a central computer.

The new era of automation, which started with the introduction of NC machine tools, was undoubtedly stimulated by the digital computer. Digital technology and computers enabled the design of more flexible automation systems, namely systems which can be adapted by programming to produce a new product in a short time. Actually, "flexibility" is the key word which characterizes the new era in automation of manufacturing systems. Today manufacturing systems are becoming more and more flexible with progress in computer technology and programming techniques.

Manufacturing systems can be divided into small stand-alone equipment, like robots and CNC machine tools, and comprehensive systems with manufacturing cells and FMSs which contain many stand-alone systems. Both types of systems are controlled either by a computer, or by a controller based on digital technology. They can accept data in the form of programs and are able to process it and provide command signals to actuators which drive slides, rotary axes, or material-handling conveyors. In stand-alone systems and simple manufacturing cells, the input data defines the position of moving slides, velocities, type of motion, etc. In more sophisticated manufacturing cells, in which a robot equipped with a vision-aid or tactile feedback device is serving a few CNC machine tools, the system makes decisions based upon the feedback signals. In FMSs the level of decisions performed by the computer is the most sophisticated in manufacturing. Parts moving on the handling conveyor are routed to the appropriate manufacturing cell by the supervisory computer. If a particular cell is busy, the computer routes the parts to another cell which is able to perform the required operations. Decisions requiring such changes in routing are accomplished in real time by the FMS computer.

The simplest manufacturing system is the NC of machine tools, such as lathes, drilling and milling machines, grinders, etc.

1-2 FUNDAMENTALS OF NUMERICAL CONTROL

Controlling a machine tool by means of a prepared program is known as numerical control, or NC. NC equipment has been defined by the Electronic Industries Association (EIA) as "A system in which actions are controlled by the direct insertion of numerical data at some point. The system must automatically interpret at least some portion of this data."

In a typical NC system the numerical data which is required for producing a part is maintained on a punched tape and is called the *part program*. The part program is arranged in the form of blocks of information, where each block contains the numerical data required to produce one segment of the workpiece. The punched tape is moved forward by one block each time the cutting of a segment is completed. The block contains, in coded form, all the information needed for processing a segment of the workpiece: the segment length, its cutting speed, feed, etc. Dimensional information (length, width, and radii of circles) and the contour form (linear, circular, or other) are taken from an engineering drawing. Dimensions are given separately for each axis of motion (X, Y, etc.). Cutting speed, feedrate, and auxiliary functions (coolant on and off, spindle direction, clamp, gear changes, etc.) are programmed according to surface finish and tolerance requirements.

Compared with a conventional machine tool, the NC system replaces the manual actions of the operator. In conventional machining a part is produced by moving a cutting tool along a workpiece by means of handwheels, which are guided by an operator. Contour cuttings are performed by an expert operator by sight. On the other hand, operators of NC machine tools need not be skilled machinists. They only have to monitor the operation of the machine, operate the tape reader, and usually replace the workpiece. All thinking operations that were formerly done by the operator are now contained in the part program. However, since the operator works with a sophisticated and expensive system, intelligence, clear thinking, and especially good judgment are essential qualifications of a good NC operator.

Preparing the part program for a NC machine tool requires a *part programmer*. The part programmer must possess knowledge and experience in mechanical engineering fields. Knowledge of tools, cutting fluids, fixture design techniques, use of machinability data, and process engineering are all of considerable importance. Part programmers must be familiar with the function of NC machine tools and machining processes and have to decide on the optimal sequence of operations. They write the part program manually or by using a computer-assisted language, such as APT. Their program is punched on a tape by means of a perforating device, e.g., a Teletype, or with the aid of the computer.

The part dimensions are expressed in part programs by integers. Each unit corresponds to the position resolution of the axes of motion and will be referred to as the *basic length-unit (BLU)*. The BLU is also known as the "increment size" or "bit weight," and in practice it corresponds approximately to the accuracy of the NC system. To calculate the position command in NC, the actual length is divided by the BLU value. For example, to move 0.7 in in the positive X direction in a NC system with BLU = 0.001 in, the position command is X + 700.

In NC machine tools each axis of motion is equipped with a separate driving device which replaces the handwheel of the conventional machine. The driving device may be a dc motor, a hydraulic actuator, or a stepping motor. The type selected is determined mainly by the power requirements of the machine.

By *axis of motion* we mean an axis in which the cutting tool moves relative to the workpiece. This movement is achieved by the motion of the machine tool slides. The main three axes of motion will be referred to as the X, Y, and Z axes. The Z axis is perpendicular to both X and Y in order to create a right-hand coordinate system, such as shown in Fig. 1-1. A positive motion in the Z direction moves the cutting tool away

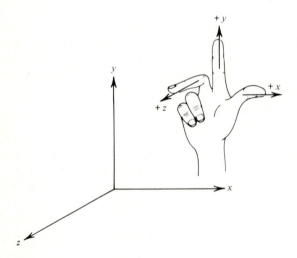

Figure 1-1 A right-hand coordinate system.

from the workpiece. The location of the origin ($X = Y = Z = 0$) may be fixed or adjustable.

Figure 1-2 shows the coordinate system of a drilling machine, a milling machine, and a lathe. In the drilling and milling machines the X and Y axes are horizontal. A positive motion command in the drill moves the X axis from left to right, the Y axis from front to back, and the Z axis toward the top. In the milling machine shown in Fig. 1-2 the directions are reversed. In the lathe only two axes are required to command the motions of the tool. Since the spindle is horizontal, the Z axis is horizontal as well. The cross axis is denoted by X. A positive position command moves the Z axis from left to right and the X axis from back to front in order to create the right-hand coordinate system.

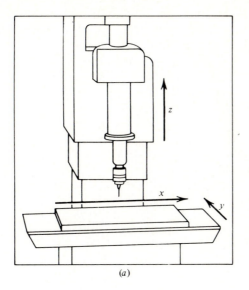

(a)

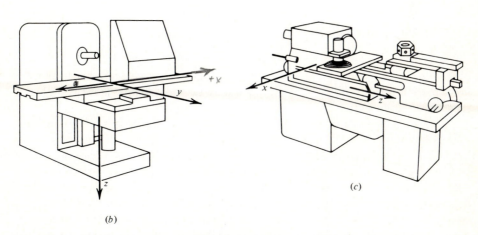

(b)

(c)

Figure 1-2 Coordinate systems in machine tools. (a) Drilling, (b) milling, (c) lathe.

If in addition to the primary slide motions secondary linear slide motions exist, they may be designated U, V, and W. Rotary motions around the axes parallel to X, Y, and Z are designated a, b, and c, respectively.

The NC machine tool system contains the machine control unit (MCU) and the machine tool itself, as is shown in Fig. 1-3. The MCU has to read and decode the part program, to provide the decoded instructions to the control loops of the machine axes

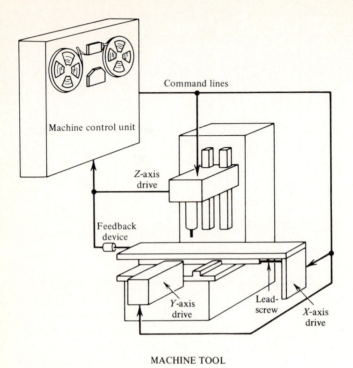

Command lines

Machine control unit

Z-axis drive

Feedback device

Y-axis drive

Lead-screw

X-axis drive

MACHINE TOOL

Figure 1-3 Numerical control system.

of motion, and to control the machine tool operation. The system also has to advance the tape each time the previous instructions were fulfilled, that is, at the end of each segment of the workpiece.

The MCU consists of two main units: the data processing unit (DPU) and the control loops unit (CLU). The function of the DPU is to decode the information received from the tape, process it, and provide data to the CLU. Such data contains the new required position of each axis, its direction of motion and velocity, and auxiliary control signals to relays. On the other hand, the CLU provides a signal announcing that the previous segment is completed and that the DPU can read a new block of the part program. The CLU operates the drives attached to the machine leadscrews and receives feedback signals on the actual position and velocity of each one of the axes. Each leadscrew is equipped with a separate driving device and a separate feedback device, but the latter exists only in a closed-loop system. The DPU includes, at least, the following functions:

Input device, such as a punched-tape reader
Reading circuits and parity checking logic
Decoding circuits for distributing data among the controlled axes
Interpolator, which supplies velocity commands between the successive points taken from the drawing

A CLU consists of the following circuits:

Position control loops for all the axes of motion (each axis has a separate control loop)
Velocity control loops
Deceleration and backlash takeup circuits
Auxiliary functions control, such as coolant on and off, gear changes, or spindle on
    and off

In CNC systems, the DPU functions are always performed by the control program contained in the CNC computer. The major part of the CLU, however, is always implemented in hardware, even in the most sophisticated CNC systems.

1-3 ADVANTAGES OF NC SYSTEMS

Before, during and especially after World War II, the U.S. Air Force increasingly felt the need to manufacture complicated and accurate aircraft parts, which were difficult to produce using conventional machine tools. The first steps in the development of a suitable machine tool were undertaken at the Parsons Company in Traverse City, Michigan, and it was completed by the Massachusetts Institute of Technology (MIT) Servomechanism Laboratory. By 1952 these research efforts had produced a NC milling machine, with three controlled axes, which is believed to be the first successful NC machine. Thus, we see that the primary motivations for the development of NC systems for machine tools were the demand for high accuracy in manufacturing of complicated parts, combined with the desire to shorten production time.

The accuracy is most important when two parts that have to be adapted one to another are produced, as for example a cylinder and a piston of a motor, and also for the manufacture of exchangeable parts, especially in the aircraft and motor industries. Producing a part that has to be cut with an accuracy of 0.01 mm or better may take a considerable amount of time using conventional methods. The operator has to stop the cutting process frequently and measure the part dimensions in order to ensure that the material is not overcut. It has been proved that the time wasted on measurements is frequently 70 to 80 percent of the total working time. NC machines save that time, while maintaining or even improving the required tolerances.

A further saving of time is achieved while passing from one operation to another during the machining of the workpiece. With a conventional machine tool, the work must be stopped at such points, since the operator has to go over to the next step. The rate of production is also decreased because of operator fatigue. In NC systems these problems do not exist at all, and, moreover, since in NC the accuracy is repeatable, inspection time is also reduced.

Contour cutting in three dimensions, or often even in two dimensions, cannot be performed by manual operation. Even when it is possible, the operator has to manipulate the two handwheels of the table simultaneously while keeping the required accuracy; thus it becomes possible only when the part is simple and requires relatively small accuracies. It is obvious that in such work the NC machine saves a considerable amount of time and improves the accuracy compared with manual operation.

It is well to keep sight of the fact that transfer lines are also designed for faster output. In these systems the machines stand in line, while the product is transferred from one machine to the other, and each machine performs a limited cycle of operations. The cycle of machining operations is simple and fixed; the process is completely automatic. Compared to NC and CNC this system has disadvantages:

1. High initial investment in the equipment.
2. A long preparation time for each production series.
3. Inflexibility of the process, since each machine is planned to make a certain fixed cycle of operation. If the part configuration is changed, the machine adjustment must be rebuilt or altered.
4. A big stock of parts is required for the process, since a part is maintained in each machine.

The transfer line is suitable only for mass production, where parts are produced in the millions. In producing small quantities, however, the machine has to be able to make many operations and change the sequence of operations in a short time. Certainly this requires a high degree of flexibility.

The feature of high flexibility is substantially significant in modern manufacturing, and especially in the aircraft industry where production is usually in small quantities. Due to rapid technological changes, a series of at most hundreds of the same model of planes are built. This means that the manufacturing machine has to be economical in use, even where a small quantity of parts is produced. This is the solution provided by NC and CNC systems: when a new product is required, only the part program has to be changed.

It is also worthwhile to note that copying machines and automatic lathes have existed for many years. The copying machine includes a stylus which moves along a master copy of the part to be produced. The stylus also has an arm which transmits the motion of the stylus to the cutting tool through an intermediate mechanism. The cutting tool then produces a part of the same shape as the master. The main disadvantage of this copying method is the time spent in producing the master copy, which is made without automation and is produced with a high degree of accuracy. With NC and CNC the master copy is not required.

The automatic lathe, which is quite flexible, has been in use for many years. The setup of the required dimensions of the parts is established by pairs of microswitches and stoppers, one pair for each segment termination. The correct placement of a microswitch establishes the dimensions and tolerance of the part. The required feed, cutting speed, and auxiliary functions are programmed, in an appropriate code, on a plugboard. The plugboard (Fig. 1-4) consists of a matrix of sockets, where information is "written" by means of plugs. Each row is referred to one segment of the workpiece. The rows are scanned in sequence, hence the method is also called *sequence control*. The appropriate limit switch provides the signal for proceeding to the next row, and the scanning is advanced in the sequence necessary to accomplish the machining of the required part.

The automatic lathe is flexible, but its main disadvantage is the adjustment pro-

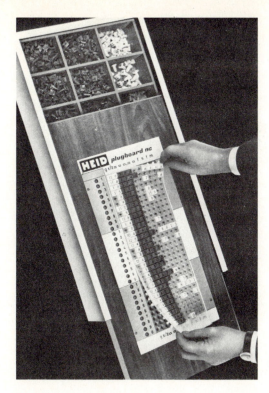

Figure 1-4 Removable patchboard for sequence control. *(Courtesy of A. G. Heid.)*

cedure of the limit switches and stoppers which consumes many working hours and requires a highly skilled and experienced operator. In addition, it should be noted that because the number of limit switches and stoppers is limited, the maximum number of programmed operations on a part is also limited. The automatic lathe can be used for a medium production series of 30 pieces or more, but it will not be a cost-effective solution for small series of say 4 or 5 pieces. On the other hand, it has been found that it might be more economical to produce even one part by NC machines than by conventional methods.

In conclusion, the NC machine tool has the following advantages, compared with other machining methods:

1. A full flexibility; a part program is needed for producing a new part.
2. Accuracy is maintained through the full range of speeds and feeds.
3. A shorter production time.
4. The possibility of manufacturing a part of complicated contour.
5. Easy adjustment of the machine, which requires less time than with other manufacturing methods.
6. The need for a highly skilled and experienced operator is avoided.
7. The operator has free time; this time may be used for looking after other machining operations.

On the other hand, the main disadvantages of NC systems are

1. A relatively high cost.
2. More complicated maintenance; a special maintenance crew is desirable.
3. A highly skilled and properly trained part programmer is needed.

Despite all of these attractive features, the penetration of NC equipment into U.S. industry has been agonizingly slow. In 1980 less than 4 percent of all machine tools in the U.S. industry were numerically controlled and accounted for only 12 percent of the total production.

1-4 CLASSIFICATION OF NC SYSTEMS

The classification of NC machine tool systems can be done in four ways:

1. According to the type of machine: Point-to-point versus contouring (continuous-path)
2. According to the structure of the controller: hardware-based NC versus CNC
3. According to the programming method: incremental versus absolute
4. According to the type of control loops: open-loop versus closed-loop

1-4.1 Point-to-Point and Contouring

Point-to-point systems. The simplest example of a point-to-point (PTP) NC machine tool is a drilling machine. In a drilling machine the workpiece is moved along the axes of motion until the center of the hole to be drilled is exactly beneath the drill. Then the drill is automatically moved toward the workpiece (with a spindle speed and feed which can be controlled or fixed), the hole is drilled, and the drill moves out in a rapid traverse feed. The workpiece moves to a new point, and the above sequence of actions is repeated. In more general terms a description of a point-to-point operation will be the following.

The workpiece is moved with respect to the cutting tool until it arrives at a numerically defined position and then the motion is stopped. The cutting tool performs the required task with the axes stationary. Upon completion of the task, the workpiece moves to the next point and the cycle is repeated.

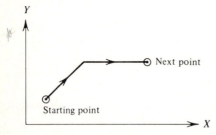

Figure 1-5 Cutter path between two holes in a point-to-point system.

In a PTP system, the path of the cutting tool and its feedrate while traveling from one point to the next are without any significance. Therefore, this system would require only position counters for controlling the final position of the tool upon reaching the point to be drilled. The path from the starting point to the final position is not controlled, as is illustrated in Fig. 1-5. The data for each desired position is given by coordinate values, and the resolution depends on the system's BLU.

Example 1-1 The *XY* table of a drilling machine has to move from the point (1, 1) to the point (6, 3); dimensions are given in inches. Each axis can move at a constant velocity of 30 in/min. Find the traveling time between the points.

SOLUTION The traveling time of the *X* axis is $(6-1)\times60/30=10$ s.
The traveling time of the *Y* axis is $(3-1)\times60/30=4$ s.
Since the axes can be moved simultaneously, the traveling time of the table is determined by the longer one, namely 10 s. The path of the table relative to the tool is similar to the one presented in Fig. 1-5.

Example 1-2 If the system's resolution is BLU=0.0001 in, what are the incremental position commands to the NC system?

SOLUTION In the *X* direction: $(6-1)/0.0001=50,000$.
In the *Y* direction: $(3-1)/0.0001=20,000$.
The commands are X+50000, Y+20000.

Contouring systems. In contouring, or continuous-path, systems, the tool is cutting while the axes of motion are moving, as, for example, in a milling machine. All axes of motion might move simultaneously, each at a different velocity. When a nonlinear path is required, the axial velocity changes, even within the segment. For example, cutting a circular contour requires a sine rate change in one axis, while the velocity of the other axis is changed at a cosine rate.

In contouring machines, the position of the cutting tool at the end of each segment together with the ratio between the axial velocities determines the desired contour of the part, and at the same time the resultant feed also affects the surface finish. Since,

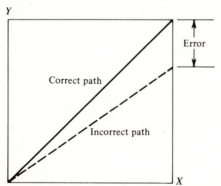

Figure 1-6 A velocity error causes position error in a contouring system.

in this case, a velocity error in one axis causes a cutter path position error (see Fig. 1-6), the system has to contain continuous-position control loops in addition to the position counters. Each axis of motion is equipped with a separate position loop and counter. Dimensional information is given on the tape separately for each axis and is fed through the DPU to the appropriate position counter. The programmed feedrate, however, is that of the contour and has to be processed by the DPU in order to provide the proper velocity commands for each axis. This is done by means of an interpolator, which is contained in the DPU of contouring systems. The function of the interpolator is to obtain intermediate points lying between those taken from the drawing.

To illustrate the interpolator function, consider a two-axis system, where a straight cut is to be made along the path of Fig. 1-6. Assume that the X axis must move p units at the same time that the Y axis moves q units. The contour formed by the axis movement has to be cut with a feedrate of V length-units per second. The numerical data of p, q, and V is programmed on the tape and is fed into the interpolator. The interpolator will then provide two velocity signals V_x and V_y, usually in a pulsed form, proportional to

$$V_x = \frac{pV}{\sqrt{p^2 + q^2}} \tag{1-1}$$

and

$$V_y = \frac{qV}{\sqrt{p^2 + q^2}} \tag{1-2}$$

In NC and CNC systems, three types of interpolators exist: linear, circular, and parabolic, but the most common in use are the linear and circular ones.

Example 1-3 A NC milling machine has to cut a slot located between the points (0, 0) and (7.1, 7.1) on the XY plane (dimensions are given in inches). The recommended feedrate along the slot is 6 in/min. Find the approximate cutting time and the axial velocities.

SOLUTION The distance traveled is $L = \sqrt{\Delta X^2 + \Delta Y^2} = 10.041$ in. The approximate cutting time is $60 (L/6) = 100.4$ s. The velocities along the X and Y axes are equal:

$$V_X = V_y = \frac{6}{\sqrt{2}} = 4.243 \text{ in/min}$$

Example 1-4 The volocity of the Y axis in Example 1-3 is off by -10 percent. What is the position error along the Y axis at the end of the path?

SOLUTION The axial velocities are $V_x = 4.243$ in/min; $V_y = 0.9V_x = 3.818$ in/min. The X axis finishes the motion in 100.4 s. During this period the Y axis travels only $L_y = tV_y/60 = 6.389$ in. The error is $7.1 - L_y = 0.711$ in. A velocity error of 10 percent causes a position error of 10 percent.

Straight-cut systems. Some PTP machines are equipped with milling capabilities also. This leads us to a third type of control, in which a contouring control is done from point to point. This is called a straight-cut system since the cutting tool can move only along straight lines which are parallel to the main axes of motion of the machine tool, as for example in a shaping machine. Cutting of the workpiece is done while the cutting tool is moving, but the latter can move along either the X, Y, or Z axis. In a straight-cut system the feedrate is programmed on the tape, and may be selected by the programmer. In this system control loops are used, but they are relatively simple. They include position counters and primitive velocity control to guarantee surface finish quality. For the latter, an error of up to ± 5 percent from the programmed feed is still permitted. An interpolator is not required for the straight-cut system since no simultaneous operation of the axes is required.

1-4.2 NC and CNC

The NC systems which were produced through the sixties used electronic hardware based upon digital circuit technology. The CNC systems, which were introduced in the early seventies, employ a minicomputer or microcomputer to control the machine tool and eliminate, as far as possible, additional hardware circuits in the control cabinet. The trend away from hardware-based NC to software-based equipment brings an increase in system flexibility and an improvement in the possibility for correcting part programs using the CNC computer.

The digital controller in hardware-based NC systems employs voltage pulses, where each pulse causes a motion of 1 BLU in the corresponding axis. In these systems a pulse is equivalent to 1 BLU:

$$\text{Pulse} \equiv \text{BLU} \qquad (1\text{-}3)$$

These pulses can actuate stepping motors in open-loop control, or dc servomotors in closed-loop control. The number of pulses transmitted to each axis is equal to the required incremental motion, and the frequency of these pulses represents the axis velocity.

In the computer the information is arranged, manipulated, and stored in the form of binary words. Each word consists of a fixed number of bits, the most popular being the 8-bit and 16-bit words. In the CNC computer each bit *(bi*nary digi*t)* represents 1 BLU:

$$\text{Bit} \equiv \text{BLU} \qquad (1\text{-}4)$$

Thus, a 16-bit word can, for example, represent up to $2^{16} = 65,536$ different axial positions (including zero). If the system resolution is, for example, BLU = 0.01 mm, this number represents motions up to 655.35 mm.

CNC systems can be designed in different configurations. The simplest one, denoted the reference-pulse approach, emulates the hardware-based NC and transmits output pulses as well. Therefore, in this system, (1-3) and (1-4) can be combined:

$$\text{Bit} \equiv \text{pulse} \equiv \text{BLU} \qquad (1\text{-}5)$$

In other CNC configurations binary words are transmitted as output. However, the actual position in these systems is measured by a digital device which transmits pulses representing BLUs. Therefore, actually in all CNC systems the terms bit, pulse, and BLU become equivalent.

The main difference in the operation between NC and CNC is in the way that the punched tape is read. In NC the punched tape is moved forward by one block and read each time the cutting of a segment is completed. During the production of each part, the tape is read again. In most CNCs the complete tape is read one time only and stored in the computer memory before the cutting starts. During the machining the control program of the CNC uses the stored part program to command the machine. By this method, tape reading errors are eliminated in CNC.

Example 1-5 The required axial positions are stored in CNC systems in software counters contained in the control program. If the maximum allowable position in a system is 9.999 in and the BLU = 0.001 in, how many counter bits are required?

SOLUTION:

$$2^n > \frac{9.999}{0.001}$$

$$n > \frac{\log 9999}{\log 2} = 13.3$$

The number of bits required is 14.

1-4.3 Incremental and Absolute Systems

NC systems may be further divided into incremental and absolute systems. An incremental system is one in which the reference point to the next instruction is the endpoint of the preceding operation. Each piece of dimensional data is applied to the system as a distance increment, measured from the preceding point at which the axis of motion was present. As an example consider Fig. 1-7 where five holes have to be drilled. The distances from the zero point to the various holes are given in the figure. Distances between points are calculated, and the X axis position commands are given as follows:

$$X + 500$$
$$X + 200$$
$$X + 600$$
$$X - 300$$
$$X - 700$$
$$X - 300$$

When considering an incremental system both the programming method and the feedback device are in incremental form. A typical feedback device is the rotary encoder which provides a sequence of pulses, where each pulse represents 1 BLU.

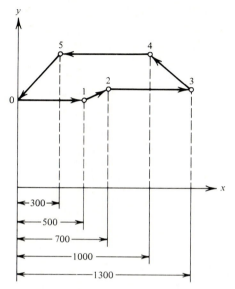

Figure 1-7 A part for drilling.

An absolute NC system is one in which all moving commands are referred to one reference point, which is the origin and is called the zero point. The position commands are given as absolute distance from that zero point. The zero point may be defined as a point outside the workpiece or at a corner of the part. If a mounting fixture is used, it could be convenient to select a point on the fixture as the zero point. In the example of Fig. 1-7, the X dimensions in the part program are written as

$$X + 500$$
$$X + 700$$
$$X + 1300$$
$$X + 1000$$
$$X + 300$$
$$X \quad 0$$

The zero point may be a floating or a fixed point. A *zero floating point* allows the operator, by pushing a button, to select arbitrarily any point within the limits of the machine tool table as the zero reference point. The control unit retains no information on the location of any previous zeros. The zero floating point permits the operator to locate quickly a fixture anywhere on the table of the NC machine.

As a matter of fact, absolute systems may be subdivided into pure absolute and absolute programming systems. By the term *pure absolute* we mean a system in which both programmed dimensions and feedback signal are referred to a single point. It therefore requires a feedback device which produces position information in absolute form, e.g., a multichannel digital encoder (see Fig. 4-8). Since this device is expensive, the pure absolute system is mainly used for rotary tables which require precise positioning control. Most of the so-called absolute systems are not equipped with an absolute feedback device but with an incremental measuring device, like an incremental encoder, interfaced with a pulse counter which provides the absolute position in BLUs. Since most absolute systems are of this type, henceforth the term absolute system will refer to an absolute programming NC system, in which all programmed dimensions refer to a single starting point.

The most significant advantage of the absolute system over the incremental one is in cases of interruptions that force the operator to stop the machine. Interruptions occur mainly from cutting tool breakage, but the machine might be stopped also for unprogrammed checking, etc. In the case of an interruption, the machine table is manually moved, and the cause of the interruption is removed. For example, in the case of a tool breakage, the operator manually moves the toolholder, exchanges the tool, and has to return the table to the beginning of the segment in which the interruption occurred. With the absolute system, the cutting tool is automatically returned to this position, since it always moves to the absolute coordinate called for, and the machining proceeds from the same block where it was interrupted. With an incremental system, it is impossible to bring the tool manually, precisely to the beginning of the segment in which the interruption occurred. Therefore, with an incremental system, each time the work is interrupted the operator must restart the part program and repeat the entire operation prior to the interruption point.

A further advantage of the absolute system is the possibility of easily changing the dimensional data in the part program whenever required. Since distances are taken from one reference point, a modification or addition of a position instruction does not affect the rest of the part program. In the case of the incremental system, the part must be reprogrammed from the point at which the original program has been modified.

Incremental systems, however, have several advantages over absolute ones:

1. If manual programming is used, with incremental systems the inspection of the part program, before punching the tape, is easy. Since the endpoint, when machining a part, is identical to the starting point, the sum of the position commands (for each axis separately) must be zero. For example, the sum of the position increments given for Fig. 1-7, is zero. A nonzero sum indicates that an error exists. Such an inspection is impossible in an absolute programming system.
2. The performance of the incremental system can be checked by a closed-loop tape. This is a diagnostic punched tape which tests the various operations and performance of the NC machine. The last position command on the tape causes the table to return to the initial position. The return of the table to its initial position is a sufficient test for normal operation of the equipment. This test is carried out at least once a day. Similar tests cannot be performed on absolute systems.
3. Mirror-image programming is facilitated with incremental systems. Mirror image in manufacturing is related to symmetrical geometry of the part in one or two axes. In this case, with incremental programming, the sign of the corresponding position commands is changed from + to − . No new calculation is required for the positions. Such a procedure in absolute systems requires a variable selection of the

zero point, which is impractical, and therefore full programming of the part is required.

Most modern CNC systems permit application of both incremental and absolute programming methods. Even within a specific part program the method can be changed. The last instruction is always programmed in the absolute method to guarantee the return of the tool to the starting point. These CNC systems provide the user with the combined advantages of both methods.

1-4.4 Open-Loop and Closed-Loop Systems

Every control system, including NC systems, may be designed as an open- or a closed-loop control. The term open-loop control means that there is no feedback, and the action of the controller has no information about the effect of the signals that it produces.

The open-loop NC systems are of digital type and use stepping motors for driving the slides. A stepping motor is a device whose output shaft rotates through a fixed angle in response to an input pulse. The stepping motors are the simplest way for converting electrical pulses into proportional movement, and they provide a relatively cheap solution to the control problem. Since there is no feedback from the slide position, the system accuracy is solely a function of the motor's ability to step through the exact number of steps which is provided at its input.

Figure 1-8 shows an open-loop and a closed-loop digital control for a single axis

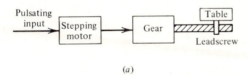

(a)

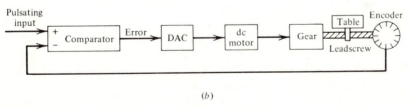

(b)

Figure 1-8 Open-loop (*a*) and closed-loop (*b*) digital control.

of motion. The closed-loop control measures the actual position and velocity of the axis and compares them with the desired references. The difference between the actual and the desired values is the error. The control is designed in such a way as to eliminate, or reduce, to a minimum, the error, namely the system is of a negative-feedback type.

In NC systems both the input to the control loop and the feedback signals may be a sequence of pulses, each pulse representing a BLU unit, e.g., 0.01 mm. The digital comparator correlates the two sequences and gives, by means of a digital-to-analog converter (DAC), a signal representing the position error of the system, which is used to drive the dc servomotor. The feedback device, which is an incremental encoder in Fig. 1-8, is mounted on the other end of the leadscrew and supplies a pulsating output. The incremental encoder consists of a rotating disk divided into segments, which are alternately opaque and transparent. A photocell and a lamp are placed on both sides of the disk. When the disk rotates, each change in light intensity falling on the photocell provides an output pulse. The rate of pulses per minute provided by the encoder is proportional to the revolutions per minute of the leadscrew.

Example 1-6 A stepping motor of 200 steps per revolution is mounted on the leadscrew of a drilling machine. The leadscrew pitch is 0.1 in. (*Note:* The pitch is the distance between successive screw threads. For a single start screw, the pitch is equal to the axial distance traveled per revolution of the screw.)
(*a*) What is the BLU of the system?
(*b*) If the motor receives a pulse frequency of 2000 pulses per second (pps), what is the linear velocity in inches per minute?

SOLUTION
(*a*) BLU = 0.1/200 = 0.0005 in
(*b*) $V = 2000 \times 0.0005 \times 60 = 60$ in/min

Example 1-7 A dc servomotor is coupled directly to a leadscrew which drives the table of an NC machine tool. A digital encoder, which emits 500 pulses per revolution, is mounted on the other end of the leadscrew. If the leadscrew pitch is 5 mm and the motor rotates at 600 rpm, calculate
(*a*) The linear velocity of the table
(*b*) The BLU of the NC system
(*c*) The frequency of pulses transmitted by the encoder

SOLUTION
(*a*) $V = 600 \times 5 = 3000$ mm/min $= 3$m/min
(*b*) BLU = 5/500 = 0.01 mm
For a motion of 1 BLU, one pulse is transmitted by the encoder.
(*c*) $F = (3000/60)/0.01 = 5000$ pps

One of the major properties of a stepping motor is that its maximum velocity depends upon the load torque. The higher the torque, the smaller the maximum allowable frequency to the motor. Stepping motors cannot be applied to machines with variable load torques, since an unpredictably large load causes the motor to lose steps and a subsequent position error occurs. In contouring systems for machine tools, the cutting forces load the motors with torques depending on the cutting conditions, and therefore stepping motors are not recommended as drives for these contouring systems. They can be applied to laser-beam contour-cutting systems (in which only a mirror is moved) and to PTP drilling machines, where the loading torque on the motors is almost constant. Industrial robots and contouring systems such as lathes or milling machines require closed-loop control systems.

Extreme care must be taken during the design of a closed-loop control system. By increasing the magnitude of the feedback signal (more pulses per one revolution of the leadscrew) the loop is made more sensitive. That is known as increasing the open-loop gain. Increasing the open-loop gain excessively may cause the closed-loop system to become unstable, which obviously should be avoided.

The design of the control system and the choice of the types of loop employed to meet performance and cost specifications, require a knowledge of the nature of the controlled machine and loading torques. The allowable positioning error, accuracy, repeatability, and response time also have to be taken into consideration where an optimum performance is required.

1-5 THE PUNCHED TAPE

In any NC or CNC system, the part program is maintained in a storage device, which can be a punched tape, magnetic tape, floppy disk, or the computer memory in CNC systems. In many NC and most CNC systems there are provisions for inserting data manually as well, which is mainly used for program debugging and setting up of the machine.

The most popular device for part program storage in both NC and CNC is the punched tape. The tape may be a paper or a plastic tape, where the latter provides more protection against damage due to chips, dirt, grease, etc. The tape is 1 in wide, with eight tracks of data punched along it, and an additional track of sprocket holes to guide the tape. Three tracks are placed on one side of the sprocket track and five on the other side, as is shown in Fig. 1-9. This unsymmetrical placement of tracks reduces the possibility of incorrect insertion of the tape in the tape reader. There can be a maximum of eight holes in a row. Such a row of holes, representing a decimal digit, algebraic sign, or letter, is called a *character*. A set of characters, which represents an instruction or any complete piece of information, is called a *word* in NC.

The instructions and data are arranged in *blocks* along the tape. Each block contains the instructions required for a specific machine movement. A block is terminated by a special end-of-block (EB) code. The information within the blocks is punched in a specific format. There are three formats in use: the tab sequential format, the word address format, and the fixed block format. In the tab sequential format, each word in a block (except the last one) is terminated by a special tab code. By counting the number of the tab codes, the control can identify a specific word in the block. The word address format uses letters to identify words in the block.

The information is punched on the tape in a standard code. Two standard codes are in use: the International Standard Organization (ISO) code, which is identical to the ASCII code, and the Electronic Industries Association (EIA) code, given in standards nos. RS-244 and RS-273. The EIA code, which is the more popular one, is characterized by an odd number of holes in each character, while in the ISO code the number of holes in a character is always an even number. It is worthwhile to note that in the

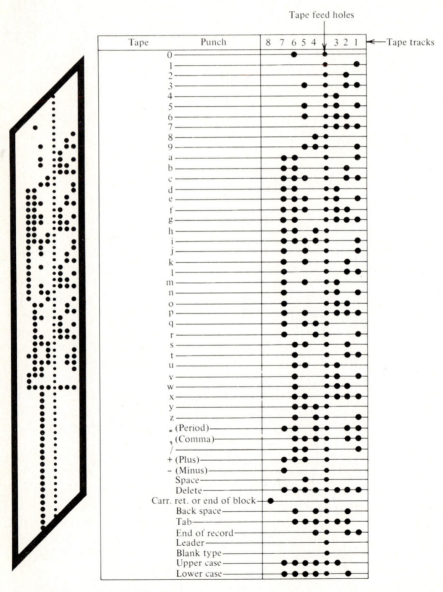

Figure 1-9 A punched tape (the EIA code).

EIA code, the EB punched code is a single hole in the eighth track and that only in this case is the eighth track punched.

The information punched on the tape is inserted into the NC system by means of a tape reader. Simultaneously with the reading of the tape, a parity check is performed for each character. That is, the system checks if the number of holes punched in a character is always an odd number for the EIA code or an even number in the cases where systems are using the ISO code. A hole which is not clear, resulting in the reading of an even number of holes in the character for the EIA code, would cause the machine to stop automatically. Similarly an odd number will stop the machine when the ISO code is used.

The tape reader reads the characters successively until the end of a block. The EB code signifies that the reading of the block is terminated, and the system has to begin performing the instructions that have just been read. The NC system performs the required segment and then sends an instruction to the tape reader to read the next block. In this manner the machining of the part proceeds: read a block, perform its instructions, and so on, until the termination of the part. In more sophisticated NC systems the tape-reading time is saved by providing a buffer storage. Simultaneously with the machining of each block, the next block is read and stored in this buffer. At the end of the block, the stored data is transferred to the appropriate units in the MCU, and the machining immediately commences. With this method, however, the reading of the complete tape for each produced part is still required.

By contrast, in most CNC systems the punched tape is read only once and then stored in the computer memory. When machining a part, the computer provides the part program to the control program in a format similar to that from the tape reader, but without pauses between the blocks. In these CNC systems, the punched tape is used only once per manufacturing series, thus avoiding tape reading errors, which are the biggest source of errors in NC systems.

The rate at which the tape is read depends on the type of tape reader. A mechanical tape reader is capable of reading about 30 characters per second. Optical tape readers have a much higher reading speed, on the order of 300 characters per second, or more. The mechanical tape reader is equipped with eight switches that are closed whenever a hole appears on the tape. The optical tape reader makes use of eight light beams and photocells which feed signals to eight lines whenever a hole is passed through the light beam.

The punched tape is prepared manually or with the aid of a computer. Manually, the tape is prepared by a Flexowriter or by a Teletype (see Fig. 1-10). These machines are electric typewriters equipped with a tape perforator and reader. They are operated like any other typewriter; however, the depression of each key of the keyboard types a character and simultaneously punches a row of holes on the tape. Both the Flexowriter and Teletype contain facilities for reproducing a tape. This is done by inserting the original punched tape in the tape reader of the equipment, and as the tape reader reads the tape, the tape perforator reproduces a new tape.

Although the punched tape continues to be the most popular device even in new CNC systems, it will probably be replaced by floppy disks in the near future.

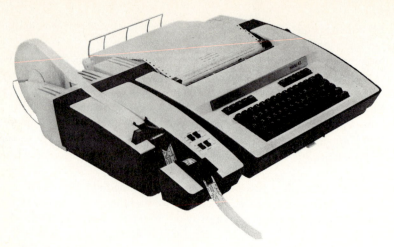

Figure 1-10 Teleprinter model 43. *(Courtesy of Teletype Co.)*

BIBLIOGRAPHY

1. Barash, M. M.: The Future of Numerical Control, *Mech. Eng.*, vol. 101, no. 9, pp. 26–31, 1979.
2. Bollinger, J. G.: Computer Control of Machine Tools, *CIRP Ann.* vol. 21, no. 2, 1972.
3. Childs, J. J.: "Principles of Numerical Control," Industrial Press, Inc., New York, 1965.
4. Cook, N. H.: Computer-Managed Parts Manufacture, *Sci. Am.*, vol. 232, no. 2, pp. 22–29, 1975.
5. Groover, M. P.: "Automation, Production Systems and Computer-Aided Manufacturing," Prentice-Hall, Inc., Englewood Cliffs, N. J., 1980.
6. Halacy, D. S.: "Computers—The Machines We Think With," Dell Publishing, New York, 1962.
7. Koren, Y., and J. Ben-Uri: "Numerical Control of Machine Tools," Khanna Publishers, Delhi, 1978.
8. Olesten, N. O.: "Numerical Control," Wiley-Interscience, New York, 1970.
9. Pressman, R. S., and J. E. Williams: "Numerical Control and Computer-Aided Manufacturing," John Wiley & Sons, Inc., New York, 1977.
10. Puckle, O. S., and J. R. Arrowsmith: "Introduction to NC of Machine Tools," Chapman & Hall, Ltd., London, 1964.
11. Roger, A.: The Microcomputer Invades the Production Line, *IEEE Spectrum*, pp. 53–57, January, 1979.
12. Wilson, F. W.: "Numerical Control in Manufacture," McGraw-Hill Book Company, New York, 1963.

PROBLEMS

1-1 The MCU of a drilling machine generates 20,000 command pulses in 12 s for controlling a stepping motor in one of the axes of motion. If the system resolution is BLU = 0.01 mm, what is the distance traveled and the velocity (in meters per minute and inches per minute) of the corresponding axis?

1-2 The MCU controls simultaneously two perpendicular axes of motion under the same conditions as in Prob. 1-1. Calculate the distance traveled and the velocity along the path.

1-3 A milling machine cuts a slot 2 in long and at 45° in the *XY* plane. The system resolution is BLU = 0.0005 in, and the required feedrate along the slot is 4 in/min. Find the distance traveled (in BLUs) and the velocity (in inches per minute) of each individual axis.

1-4 If the velocity of the Y axis in Prob. 1-3 decreases by 0.5 percent, what is the maximum contour error (in BLUs) in the Y direction? Assume that motion in each axis is completed after 2828 BLUs; the path is shown in Fig. 1-6.

1-5 Would an open-loop control be suitable for a feed drive system of

(a) A drilling machine table which always moves at the same velocity and has the same load?

(b) A milling machine which moves at various velocities with different loads?

Explain your answer.

1-6 Design an open-loop control system using a 200-steps-per-revolution stepping motor as the axial drive device; the required resolution is BLU = 0.01 mm.

(a) What should be the pitch of the leadscrew if it is directly coupled to the motor? Assume a single start screw.

(b) For a given leadscrew of 5-mm pitch, what is the required gear ratio between the motor and the leadscrew?

• **1-7** Select a stepping motor for a NC drive. Given the leadscrew pitch = 0.2 in, the resolution BLU = 0.001 in, and the maximum required velocity 20 in/min. Note that to select a stepping motor means to determine the number of required steps per revolution and the maximum allowable frequency to the motor.

1-8 Derive a general formula for open-loop control, which relates the system resolution (BLU), the leadscrew pitch (P), and the gear ratio between the motor and screw (K_g) to the required steps per revolution (M) of the stepping motor.

1-9 In a contouring system a dc servomotor is mounted to the leadscrew through a gear ratio of 2:1. When the motor rotates at 800 rpm, the axial velocity is 80 in/min. Find the leadscrew pitch.

• **1-10** A closed-loop NC system applies an incremental encoder as the feedback device. If the leadscrew pitch is 0.1 in and the system resolution is BLU = 0.0001 in, what is the encoder gain? Note that "encoder gain" means the number of transmitted pulses per revolution of the encoder.

1-11 Design a closed-loop control system using a dc motor as the axial drive element and an incremental encoder as the feedback device. The maximum speed of the motor is 1800 rpm. The system is equipped with a leadscrew of 10-mm pitch. The maximum required feedrate is 6 m/min and the required resolution is 0.01 mm.

(a) What is the encoder gain?

(b) What is the required gear ratio?

(c) What is the maximum frequency of pulses emitted by the encoder?

1-12 Derive a general formula which relates the dc servomotor speed (N), the gear ratio (K_g), and the leadscrew pitch (P) to the axial velocity (V).

TWO

FEATURES OF NC MACHINE TOOLS

Numerically controlled machine tools must be better designed, better constructed, and more accurate than conventional machine tools. Since NC machines require a sizable capital investment, an efficient usage of the equipment becomes a necessity. It is therefore necessary to minimize all noncutting machine time, apply fast tool changing methods, and minimize idle motions by increasing the rapid traverse velocities.

2-1 FUNDAMENTALS OF MACHINING

Machining is the manufacturing process in which the size, shape, or surface properties of a part are changed by removing the excess material. Therefore, machining is a relatively expensive process that should be specified only when high *accuracy* and good *surface finish* are required, which, unfortunately, are necessary for almost all mating parts.

There are five basic types of machine tools to perform machining: lathe or turning machine, drilling or boring machine, milling machine, shaper or planer, and grinder. The first four, referred to as the basic machine tools, are similar in that they use cutting tools that are sharpened to a predetermined shape, whereas, the cutting edges on a grinding wheel are not controlled.

The machining operation is a chip formation process which is accomplished through the relative motion of a tool and the workpiece, as is done with machine tools. However, the material removal mechanism may be chemical, electrical, or thermal as in the case of the unconventional processes such as electrochemical machining (ECM), electrical discharge machining (EDM), and laser-beam machining.

The cutting conditions in machining with NC machines usually refer to those variables that are changed by the part programmer and that affect the rate of metal removal. They are the cutting speed and the size of cut, which refers to the feed and depth.

The *cutting speed v* is defined as the relative velocity between the cutting tool and the work material, and is expressed in units of feet per minute, or meters per minute. In some machines (e.g., lathes) it is the work that rotates to provide the cutting speed, whereas in other machine tools (e.g., drilling and milling machines) it is the cutter that moves to provide the cutting speed. In NC machines the *spindle speed,* rather than the cutting speed, is programmed. The spindle speed is calculated by the part programmer based on the required cutting speed and diameter of the workpiece or tool.

The *depth of cut d* is defined as the distance the cutting tool projects below the original surface of the work and is expressed in thousandths of an inch or in millimeters. The depth of cut determines one of the linear dimensions of the cross-sectional area of the size of cut. The *feed f* is the second linear dimension that determines the cross-sectional area of the size of the cut. The feed is defined as the relative lateral movement between the tool and the workpiece during a machining operation. On the lathe and drill it is expressed in length-units per revolution: inches per revolution or millimeters per revolution. On the milling machine it is expressed in length-units per tooth: inches per tooth or millimeters per tooth. NC machines, however, are programmed with the units length per minute (inches per minute or millimeters per minute), which is denoted as *feedrate*. In milling operations the feedrate is the product of the basic feed times the number of teeth in the cutter times the revolutions per minute of the cutter. In turning, the feedrate is the product of the feed and the spindle revolutions per minute.

The product of the proper speed, feed, and depth of cut determines the *metal removal rate (MRR),* which is expressed in volume-units per minute. The productivity of the machine during cutting is proportional to the MRR.

Example 2-1 A cylinder of 6.1 in diameter is to be reduced to 5.9 in diameter in one turning cut with a feed of 0.006 in/r and a cutting speed of 500 ft/min on a NC lathe. Calculate the following:
(*a*) The programmed spindle speed in revolutions per minute
(*b*) The programmed feedrate
(*c*) The MRR

SOLUTION
(*a*) When the cutting speed is expressed in feet per minute, the spindle speed is

$$N = \frac{12v}{\pi D} \tag{2-1}$$

where D is the average diameter in inches. On the basis of Eq. (2-1), the spindle speed is 318 rpm.
(*b*) The feedrate is the product of the feed and the spindle speed:

$$F = fN \tag{2-2}$$

which gives $F = 1.9$ in/min.

(c) The MRR is calculated by

$$\text{MRR} = 12vfd \tag{2-3}$$

The depth of cut d is one-half of the decrease in diameter. MRR $= 3.6$ in^3/min.

When SI units are used, the cutting speed is given in meters per minute, the feed in millimeters per revolution, and the depth of cut and workpiece diameter in millimeters. Therefore in SI units the spindle speed is calculated by

$$N = \frac{1000v}{\pi D} \tag{2-4}$$

and the feedrate F is given in millimeters per minute. The values of N and F are used by the part programmer. The equation of MRR given in cubic millimeters per minute is

$$\text{MRR} = 1000vfd \tag{2-5}$$

Example 2-2 A slot is machined on a milling machine with a two-tooth end mill cutter at a feed of 0.005 inches per tooth. The spindle rotates at 1000 rpm. What is the machining feedrate used by the part programmer?

SOLUTION In milling operations, the feedrate is calculated by

$$F = pfN \tag{2-6}$$

where p is the number of teeth in the cutter. For the above data, Eq. (2-6) yields $F = 10$ in/min.

2-2 DESIGN CONSIDERATIONS OF NC MACHINE TOOLS

The reasons that stimulated the development of NC were the demand for better *accuracy* in the manufacturing of complicated parts and the desire to increase *productivity*. Digital control techniques and computers have undoubtedly contributed toward achieving these goals. However, it should be noted that the combined characteristics of the control and the machine tool determine the final accuracy and productivity of the NC or CNC system.

The term accuracy is often mistakenly interchanged with the terms resolution and repeatability. The *resolution* of an NC or CNC system is a feature determined by the designer of the control unit and is mainly dependent on the position feedback sensor. One has to distinguish between the programming resolution and the control resolution. The programming resolution is the smallest allowable position increment in part programs and is referred to as the BLU which might be on the order of 0.01 mm in a typical machine tool system. The control resolution is the smallest change in position that the feedback device can sense. For example, assume that an optical encoder which emits

1000 voltage pulses per revolution of the shaft is directly attached to a 10-mm-pitch leadscrew on a machine tool table. This encoder will emit one pulse for each 0.01 mm (10/1000) of linear displacement of the table. The unit 0.01 mm is the control resolution of this system. Displacements smaller than 0.01 mm cannot be detected. To obtain the best system's efficiency, the programming resolution is equal to the control resolution and is denoted as the system resolution, or BLU.

The final accuracy of a CNC system depends on the computer control algorithms, the system resolution, and the machine inaccuracies (which are discussed below). Control algorithms might cause position errors due to round-off errors in the computer. In machine tools it might occur, for example, during the interpolation in circular motions (further discussed in Chap. 5). These types of errors, however, affect only the contour accuracy and do not cause position errors at the end of the segments of the machined part.

System inaccuracy due to resolution is usually considered to be $\frac{1}{2}$ BLU. The reason is that displacements smaller than 1 BLU can be neither programmed nor measured and, on the average, they count for $\frac{1}{2}$ BLU. When the inaccuracies associated with the machine itself are included, a poorer accuracy will result. Therefore, the following relationship can be used to determine a realistic system accuracy:

$$\text{System accuracy} = \tfrac{1}{2} \text{ BLU} + \text{machine accuracy}$$

The machine designer tries to ensure that the accumulated effect of all inaccuracies associated with the machine tool will be under $\frac{1}{2}$ BLU, and then the system accuracy becomes approximately equal to the system resolution.

Repeatability is a statistical term associated with accuracy. If a machine slide is instructed to move from a certain point the same distance a number of times, all with equal environmental conditions, it will be found that the resultant motions lead to discordant displacements. The system repeatability is the positional deviation from the average of these displacements. It is clear that through a greater number of such experiments we will be able to give a more precise estimate of the repeatability. The repeatability will always be better than the accuracy.

High productivity and accuracy might be contradictory features. High productivity calls for higher speed, feed and depth of cut, which increase the heat and cutting forces in the system. This might lead to thermal deformations, deflections, and vibrations of the machine, and consequently the accuracy deteriorates. Therefore, the structure of the NC machine tool must be more rigid than that of its conventional counterpart.

One design feature common to all machine tools is the materials they are made of. The conventional machine tools in the past were made of cast iron, and the better ones had the slide surfaces flame-hardened. NC machines, however, are usually all-steel-welded constructed machines in both the light-duty and heavy-duty sizes. The advantage of a welded steel structure over a cast iron one is that greater strength and rigidity are obtained for a given weight.

In addition to the improved structure, better accuracy is obtained in NC machines by using low-friction moving components, avoiding lost motions and isolating thermal sources. In regular slides the static friction is generally higher than the sliding friction.

Therefore, the force which is applied to overcome the static friction becomes too big when the slide starts to move. Because the slide has inertia, it goes beyond the controlled position, stops, and the cycle is repeated. This affects the accuracy and the surface finish of the part. To avoid this phenomenon, slides and leadscrews in which the static friction is lower than the sliding friction must be used.

· Lost motions mean uncontrolled motions which may cause dimensional errors in the part, which cannot be corrected by the closed-loop control system. These motions include deflection of the tool due to cutting forces, and backlash, or free play, in the axial drive mechanism. The feedback device is mounted on the leadscrew in many NC systems. In these cases, any backlash between the leadscrew and the table or the toolholder may cause an equivalent error in the part. With PTP systems this problem can be solved with the aid of digital circuits. Since the objective here is only static positioning, the designers use "backlash take-up circuits," which ensure that the final approach to the point is always from the same direction. In contouring systems, however, the machining path also includes points where an axis of motion reverses, and backlash errors are not eliminated.

Example 2-3 In an NC drive the pitch of the leadscrew is 10 mm, and an encoder of 1000 pulses per revolution is mounted on its end. The backlash between the leadscrew and the nut is 3.6°. Find the backlash in terms of linear slide movement and BLUs.

SOLUTION The linear backlash is $10 \times (3.6/360) = 0.1$ mm. The system's BLU is $10/1000 = 0.01$ mm; therefore the backlash is equivalent to 10 BLUs of linear motion.

High productivity is achieved in NC by increasing the efficiency of the machine: using machining centers and turning centers rather than milling machines and lathes. These centers permit the employment of higher feeds and depth of cut to increase the MRR, together with the usage of faster idle feedrates and automatic tool changers to shorten the nonmachining time.

Despite the high productivity and accuracy achieved with NC machine tools, NC is not always a cost-effective solution in the manufacturing of parts. If mass production of parts is required (i.e., more than 100,000 parts annually), a set of special-purpose machines arranged in a transfer line is more economical than the use of NC machines. This situation is often encountered in the automotive industry. If only several (e.g., fewer than 20) simple parts are required, production with conventional machine tools is less expensive than with NC machines. Production with NC and CNC machines is economical when relatively complex parts are produced in medium-size lots, or batches (e.g., 20 to 10,000 parts annually). This situation is often found in the aerospace industries, where there is a great variety in required parts and the quantities per type are not as large as in the automotive industry. However, even the production of a single part on an NC machine can be economically justified if the part contains complicated surfaces which make it extremely difficult to be produced on conventional machine tools.

Small machine shops and research institutes often consider retrofitting a conventional machine tool into an NC system in order to eliminate the purchase of a new machine. A system including a high-power retrofitted lathe interfaced in a closed-loop with a PDP-11/40 minicomputer is shown in Fig. 2-1. This system, which was designed at the Technion, Israel Institute of Technology, can be operated either from a control panel or via the computer keyboard.

Figure 2-1 Retrofitted CNC lathe (the project was executed at the Technion, Haifa, Israel).

Stand-alone MCUs, including a microcomputer as an integral part of the controller, are commercially available and can be applied to retrofitted machine tools. The MCU contains either translator amplifiers for stepping motor drives, or power amplifiers for dc servomotor drives. One amplifier must be provided for each axis of motion. Aside from adding an MCU, the retrofitting includes modifications to the machine itself, such as

1. Replacing the conventional leadscrews with ball-bearing screws.
2. Adding stepping motors (for open-loop control), or dc servomotors and feedback devices (for closed-loop control); the drive is added to each controlled axis.

If control is applied to three or more axes, the total expense becomes considerable and the project is not always cost-effective.

2-3 METHODS OF IMPROVING MACHINE ACCURACY

The usual procedure of controlling accuracy during conventional machining is to measure manually the dimension in question and compare it with the required one. If an error is found, manual adjustment of the machine table or the tool becomes necessary. By contrast, with NC machines the actual part dimensions are not directly measured by the system. Although a feedback device is mounted on the leadscrew and measures its precise position, the NC system does not control the position of the cutting edge of the tool, and this can cause error.

2-3.1 Tool Deflection and Chatter

The force of the tool edge against the workpiece in milling and turning deflects the tool and the toolholder and consequently causes dimensional errors in the machined part. Special care must be taken to minimize such errors, usually by increasing the stiffness of the construction of the tool mounting.

Another phenomenon which might occur from the tool deflection is a vibratory response called *chatter*. In turning, a chatter might start if during one revolution of the workpiece, the cutting tool deflects at one point more than the average, and as a result a lump is left on the workpiece at that point. When, after one revolution of the workpiece, the tool arrives again at the point with the lump, the effective load on the tool is larger, carrying a consequently greater tool deflection, which, in turn, leaves a new lump on the workpiece. Since this process is repetitive, it causes vibrations of the tool at a frequency proportional to the spindle speed. The chatter occurs as a function of the machine structure, the tool and workpiece material, and the cutting conditions. Using machine tools with greater stiffness can eliminate a chatter which occurs under the same cutting conditions on machines with less stiffness.

2-3.2 Leadscrews

Inaccuracies are also caused by the mechanical linkage between the leadscrew and the tool. This linkage actually functions in open-loop, since the position feedback device is mounted on the leadscrew. In order to improve the accuracy, this mechanism must be time-invariant (no heating effects) and linear (no backlash and friction).

The driving source of each axis of motion is mounted on the axis leadscrew. The power is transmitted to the table and toolholder head (in drilling and milling machines), or to the carriage (in a lathe), by means of a nut that engages the leadscrew. The table, or the carriage, can be moved over the base along a mating set of slides. In conventional machine tools a regular leadscrew and lubricated dovetail slides are used. The friction of these elements during motion is a regular sliding friction.

For lubricated surfaces the curve of coefficient of friction μ plotted against velocity is shown in Fig. 2-2. Two sections of it are labeled. In the zone labeled "coulomb friction" the surfaces are in contact. This gives rise to the so-called sticking forces when surfaces start to move from rest. In the zone marked "fluid film lubrication" the surfaces are separated by an hydrodynamic oil film, and the frictional force varies approximately with the sliding velocity.

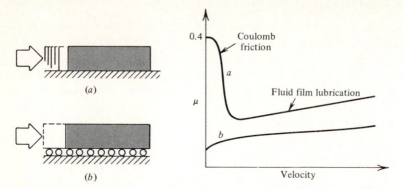

Figure 2-2 Friction behavior versus velocity. (*a*) Sliding friction, conventional screw; (*b*) rolling friction, ball-bearing screw.

On plain machine tool slides, friction is normally coulomb, after the surfaces have been at rest for a few seconds. This condition makes accurate control of the feedrate axes difficult to achieve since a considerable force may be necessary before movement commences. Once it does commence, the force necessary may drop away rapidly, thus giving rise to the well-known "stick-slip" phenomenon which causes uncontinuous motion and shudder at low velocities and consequently affects the part accuracy. This problem may be avoided by using rolling element bearings, hydrostatic bearings, or ball-bearing leadscrews.

A *ball-bearing leadscrew* is well-described by its name: it is a leadscrew that runs on bearing balls, as shown in Fig. 2-3. The screw thread is actually a hardened ball race. The nut consists of a series of bearing balls circulating in the race and carried from one end of the nut to the other by return tubes. The balls provide the only physical contact between the screw and nut, replacing the sliding friction of the conventional leadscrew with a rolling motion. The substitution of rolling contact for sliding metal-to-metal contact minimizes starting friction and eliminates the tendency for stick-slip when a slow smooth motion is desired. Notice that a slow axial velocity might be required even at high feedrate, when a combined motion of two axes is required to produce the segment, such as a conical cut on a lathe.

The high efficiency (about 90 percent) achieved with rolling contact devices permits the employment of antibacklash methods. The backlash can be somewhat reduced by loading the nut with oversized balls. To reduce the backlash to zero, however, it is necessary to use a double-nut configuration and to preload one nut against the other by using a spring or other resilient element. The preload should not exceed about one-third of the operating load to achieve an efficient leadscrew and still eliminate the backlash.

Another advantage of the ball-bearing leadscrew is the extended life of the screw compared with the conventional one. It also eliminates the need for frequent compensating adjustments for wear, since due to the rolling contact, very little dimensional change occurs over the life of the leadscrew.

Figure 2-3 Ball-bearing leadscrew. *(Courtesy of Warner Electric.)*

In very precise NC machines the dovetail slideways are replaced by roller bearing slides. Many forms have been developed for machine tool slides. A popular type is a round slide with recirculating balls running between the slide and a nut, thereby replacing the sliding with rolling friction and reducing the stick-slip problems at slow motions.

2-3.3 Thermal Deformations

Errors in NC also arise due to thermal deformation of the machine tool. There are three sources of heat: the machining process, the spindle and driving motors, and the friction of slideways and leadscrews. A nonuniform distribution of the thermal sources along the machine can cause deformations of the toolholder, carriage, table, etc. A temperature difference of 1°C along 1000 mm can cause an error of 0.01 mm, which may be within the required machine accuracy. Therefore, the machine tool manufacturer must take the thermal effects into consideration in the design stage of the machine. This includes removal of high-power motors from the machine base, providing large heat-removing surfaces, use of low-friction bearings, and symmetrical distribution of the heat sources.

The heat effect can be minimized by the machine designer, but it cannot be completely removed. Different machining conditions vary the temperature of the cutting process, which is one of the main heat sources. Therefore, machine tools which require precise machining are always located in an air-conditioned environment or separate rooms. If a higher accuracy is required, the employment of expensive special measuring devices and compensation with additional feedback loops become necessary.

2-4 INCREASING PRODUCTIVITY WITH NC MACHINES

The total production time of any machining job is comprised of four individual times:

1. Actual cutting time
2. Idle and traverse motion time
3. Loading and unloading time
4. Tool changing time

The actual cutting time is inversely proportional to the product of the cutting parameters: cutting speed, feed, and depth of cut. This time can be only slightly reduced by applying NC machines, since the cutting parameters are determined by economic analysis and permittable loading forces on the tool. However, production time can be substantially saved by reducing the other three time components mentioned above.

Idle or traverse motion means movements of the machine axes along which cutting does not occur. In a turning process about half of the motions are of this type. In milling, idle motions occur when traveling from the starting point to the workpiece and back at the beginning and end of the operation. By increasing the traverse velocity, the idle time is reduced, and production time is saved. The maximum permittable traverse velocity depends upon the rigidity of the machine, drives, leadscrews, and type of slides.

Loading and unloading time can be saved by using two part-holding fixtures simultaneously on the machine table. One part can be unloaded and the next workpiece loaded, while the other is being machined. Likewise, the same principles apply with two worktables which can be automatically switched.

A substantial saving in production time is obtained with *automatic tool-changing* methods. Tools must be changed in complex machining operations, in which many tools are used to produce a part. The number of tools needed for completing the operations depends upon the complexity of the job. For simple turning and drilling operations, six or eight tools may be adequate, and this restricted number would enable the use of a simple *turret* machine. The simple, quick motion that indexes the turret from one tool to the next makes it a fast method of tool changing. However, since in this method the cutting tool never leaves the spindle during a machining sequence, the turret is not regarded as an automatic tool changer (ATC).

An ATC is a device containing a rotating tool storage with an automatic exchange of tools at the spindle. The tools are available for automatic selection by the MCU. The tool storage is either a chain-type (Fig. 2-4) or a carousel-type magazine (Fig. 2-5). The

Figure 2-4 Chain-type tool magazine. *(Courtesy of Scharmann.)*

Figure 2-5 Carousel-type tool magazine. *(Courtesy of Giddings & Lewis.)*

magazine provides storage for the tools which are required to perform all machining operations on the part. A special mechanism is employed to select the correct tool and transfer it to the spindle. The tool that was in the spindle is removed and replaced in the storage magazine, where it can be relocated the next time it is needed.

Most ATCs of both the chain- and the carousel-type magazine have a changer arm (see Fig. 2-6) to exchange tools between the magazine and the machine spindle. At the

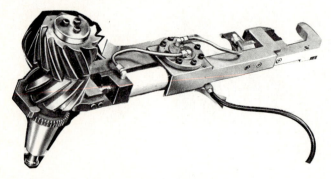

Figure 2-6 Tool changing arm of automatic tool changer.

start of each machining operation the magazine is automatically rotated, and the next tool is brought to the changing position. Tool changing involves the following steps:

1. The changer arm rotates 90° and engages the tools in the spindle and magazine simultaneously.
2. The arm grips the tools mechanically, then moves forward to remove the tools from the socket and spindle.
3. The changer arm continues the rotation with the two tools to change the position of the tools by 180°.
4. The arm retracts back to the machine column and places the selected tool in the machine spindle and the used tool in the magazine; the collet closes on the toolholder in the spindle.
5. The arm rotates an additional 90° to its rest position, while the machining operation resumes.

With this procedure the used tool is returned to the location just vacated by the newly selected tool. This requires that each toolholder be identified by some form of coding device which can be recognized by the selector mechanism. Typical coding systems for tool identification are based either on the use of groups of rings on the toolholder or on code keys related to a definite tool.

2-5 MACHINING CENTERS

The concept of a machining center is to bring the machine to the work, instead of vice versa. In the earliest applications, such items as floor plates, angle plates, and indexing tables were set down in convenient relationship to a number of machine tools (traveling-column boring and milling machines and radial drills, for example) with the object of enabling large workpieces to be machined with a minimum of handling and different setups.

Such installations occupied substantial floor areas. However, in those industries where lengthy and complex machining cycles needed to be carried out on large and heavy workpieces, these early machining centers served very efficiently, and many still are used in production.

In recent years, however, the term machining center has come to be associated almost exclusively with a single machine, usually incorporating automatic tool changing, a rotary table which facilitates circular machining, and one or two integral work-tables. Such a machine is usually designed to perform operations on at least four, and sometimes five, surfaces. There are machining centers with automatic tool changers in vertical and horizontal configurations, as shown on Figs. 2-7 and 2-8, respectively. A typical machining center may use 50 or more tools.

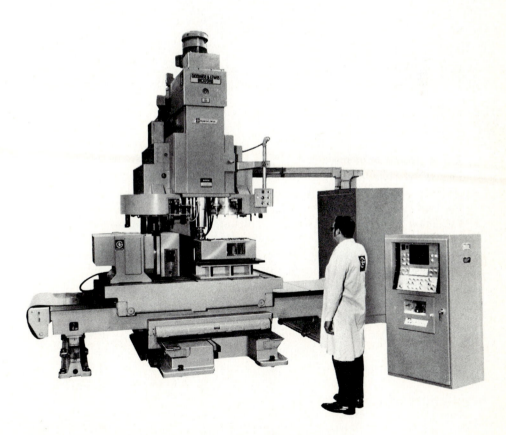

Figure 2-7 Vertical machining center. *(Courtesy of Giddings & Lewis.)*

Figure 2-8 Horizontal machining center—The Milwaukee Matic 800. *(Courtesy of Kearney & Trecker.)*

The development of machining centers has been stimulated by the need for high accuracy in the manufacturing of large and complex parts. Recent machining center designs have emphasized the following features:

1. Reduction of nonproductive time by applying faster tool changing and more rapid traverse motions
2. Increased accuracy by using rigid welded construction for beds and columns
3. Heavier allowable cutting depths and feeds
4. Improved user-oriented machine diagnostics
5. Using bubble memory to store many part programs

The early machining centers were primarily verticals, since people were trained and became familiar with vertical configurations of conventional machines. In addition, the vertical machine is normally less expensive and more flexible. The trend in the large shops is to start out with a vertical machine and then go to the horizontal machining center.

One fairly new development in horizontal machining centers is the traveling-column design, which is shown in Fig. 2-9. Instead of utilizing a pallet-changing

Figure 2-9 Horizontal machining center with traveling column. *(Courtesy of Giddings & Lewis.)*

mechanism that moves the workpiece to the spindle, the column goes to the work. Machines with a traveling column have dual tables, so the machine acts like a miniature production line with workpieces being changed on the idle table. There is also a trend toward using machining centers in transfer lines since the lines are being built with the capability of switching over from one part design to another. The moving column maintains parallelism between the machining center and the transfer line, so it fits right into the design of the line.

It is apparent that the production of simple parts is more cost effective on a two-axis turret drill than on a vertical machining center equipped with an ATC, which costs more than twice as much. However, it should be noted that the production rate for such parts is often higher on the turret drill than on the machining center, mainly because the tool changing time on an indexing turret is faster than the changing time of the ATC. The time taken to change tools becomes increasingly significant when machining small workpieces due to the shorter cutting times per tool. With a small workpiece the time required for a tool change can be even greater than the cutting time of the tool, so that the effect of the tool changing time can become considerable.

Example 2-4 Eight tools are required to machine a two-dimensional part. The total actual machining time is 150 s. Compare the production time of a two-axis, eight-station turret drill, in which the changing time is 2 s, with a vertical machining center in which 5 s is required to change tools.

SOLUTION The total tool changing time on the drilling machine is 16 s and on the machining center is 40 s. Therefore, the ratio between the production times is

$$\frac{150 + 40}{150 + 16} = 1.15$$

The production on the turret drill is 15 percent higher than on the machining center.

Analogous to machining centers are the turning centers. In general turning centers are more accurate, powerful, and flexible than conventional NC turret lathes and can perform a wider range of operations. The cutting tool is located above the workpiece, rather than in front of it, which causes the chips to fall directly down, as shown in Fig. 2-10. In this machine an automatic chip remover was added to further decrease the nonproductive time. Both machining centers and turning centers can make a substantial contribution to machining large and complex components with high accuracy, which requires a variety of machining operations.

Figure 2-10 Turning center. *(Courtesy of Giddings & Lewis.)*

2-6 MCU FUNCTIONS

The features contained on the control panel of the MCU are the direct interface between the machine operator and the NC system. The more common switches available on the controller panel are listed below.

2-6.1 Mode Selection

Several modes of operation are usually available in NC systems: auto mode, which is the normal mode; the manual or dial-in mode; the jogging mode; and the block-by-block mode.

Auto mode. This mode permits fully automatic operation of the NC system, namely, continuous execution of the part program.

Manual data input mode. This permits the operator to dial in information to the control and thus instruct the machine to follow machining cycles without the use of a punched tape; this is useful when setting up. Thumbwheel switches are used to select information such as m and g functions, spindle speeds, feeds, and coordinate dimensions, and a button is used to transfer this information into the control, one block at a time. The cycle-start button is used to execute each block of information.

Jogging switches. For setup purposes it is an advantage to be able to control slide movements. Jogging switches permit this by selecting direction and appropriate axis and then pressing a button to move the slide. The jogging mode is used for course positioning; the fine positioning is carried out by using discrete jogging switches.

Discrete jogging switches. These switches permit the axes to be moved a known distance, usually 0.1 in, 0.01 in, 0.001 in, or 0.0001 in. The axis and the increment of movement are selectable, and the slides move at the press of a button. This feature can be used in setting up and in selecting depths of cut.

Block-by-block mode. When in the block-by-block mode with the tape in position, it is possible for the control to read and execute one block of information at a time. Successive presses of the cycle-start button will permit further blocks of the part program to be read and executed block by block.

2-6.2 Compensations and Override

The following compensations and feedrate corrections are usually available.

Tool zero offsets or cutter radius compensation. This facility is usually in the form of a series of decade switches for each axis of motion. In order to offset the actual tool path from the programmed tool path, the amount of offset can be dialed in on the switches so the tape does not need to be modified.

This feature is particularly useful in milling applications where resharpened cutters, or cutters of different diameters than the programmed size, can be used without altering the tape. Roughing and finishing cuts can be made from the same tape, and modifications to programmed dimensions may also be made.

Tool zero offsets are frequently added to lathe controls to compensate for tool wear and to allow different diameters to be machined in order to be matched to mating parts.

Tool length compensation. This feature is similar to tool zero offset but applies to the spindle axis. It is particularly suitable in drilling applications where the varying lengths of drills can be compensated.

Feedrate override. It is often advantageous to override the programmed feedrate, especially in cases where adverse cutting conditions are encountered. Chilled castings, dull cutting tools, hard materials, and unsuitable programmed feedrates can be dealt with by simply adjusting the feedrate on a dial usually calibrated in percentages of the programmed feedrate. Although some manufacturers offer 0 to 140 percent which also provides for increased feedrates, 0 to 100 percent is the most common.

2-6.3 Readout Displays

Sequence number readout. Each block of information has a sequence number (e.g., n023 signifies the twenty-third block), and it is often convenient to have that number displayed so that the operator knows which block of the tape is being read at a particular time. This facility is very useful when checking tapes, because it enables the operator to know the location of any tape errors.

Present-position readout. This enables the operator to see on electronic displays the present position of one or all axes. This becomes particularly useful when the machine is being used manually and also when setting up the NC system.

2-6.4 CNC Controllers

CNC systems also contain other features associated with the computer of the CNC controller. These include a keyboard to aid in tape editing and a cathode-ray tube (CRT) screen on which messages are displayed for the operator. The availability of MCU features is quite important, since they contribute to the satisfactory performance of the entire CNC system.

BIBLIOGRAPHY

1. Astrop, A. W.: The Yamazaki Machining System, *Machinery,* February 23, 1977.
2. Datsko, J.: "Material Properties and Manufacturing Process," Joseph Datsko consultant, Ann Arbor, Mich., 1966, 7th printing, 1977.
3. Kearney and Trecker: "Milwaukee-Matic," Pub AD 121-5170, Brighton, England, 1970.
4. Koren, Y., and J. Beh-Uri: "Numerical Control of Machine Tools," Khanna Publishers, Delhi, 1978.
5. NC Turret Drill or Machining Centre, *Mach. Tool Rev.,* no. 383, pp. 66–69, May–June, 1978.
6. Simon, W.: "Numerische Steuerung von Werkzeug-Maschinen," Hanser Verlag, Berlin, 1963.
7. Vasilash, G. S.: Machining Centers on the Move, *Mfg. Eng.,* pp. 94–97, September, 1980.
8. Warner Electric: "Ball Bearing Screws & Splines," 1980.
9. Weggen, E.: Machining Centers, Today New Conception, *Scharmann, T. Z. F. Prakt. Metallbearb,* no. 113, 1976.
10. Welbourn, D. B., and J. D. Smith: "Machine-Tool Dynamics: An Introduction," Cambridge University Press, London, 1970.
11. Wilson, F. W.: "Numerical Control in Manufacturing," McGraw-Hill Book Company, New York, 1963.

PROBLEMS

2-1 A cylindrical part of 2 in diameter is machined on a lathe at a cutting speed of 600 ft/min and a feed of 0.01 in/r. Find the spindle speed and the feedrate (in inches per minute) that the programmer uses in preparing the part program.

2-2 A cylindrical part of 80 mm diameter and 250 mm length is machined at a cutting speed of 200 m/min. Two turning cuts are required: the rough cut is performed at 0.5 mm/r with a depth of cut of 4 mm and the fine one at 0.1 mm/r. Find the spindle speed, the feedrates (in millimeters per minute), and the actual machining time.

2-3 A straight cut is performed on a milling machine with a four-tooth cutter at a feed of 0.003 inch per tooth. If the spindle rotates at 900 rpm, what is the axial feedrate in inches per minute?

2-4 A retrofitted lathe is interfaced with a 16-bit computer. One bit is reserved for the sign to determine direction of motion. What is the maximum position command which can be programmed (with 15 bits) if the system resolution is BLU = 0.001 in.

Greg   12:00   Tuesday

THREE

NC PART PROGRAMMING

The conversion of the engineering blueprint to a part program can be performed manually or with the assistance of a high-level computer language. In both cases part programmers determine the cutting parameters, such as spindle speed and feed, based upon characteristics of the workpiece, tool material, and limitations of the machine tool. Therefore they must have extensive knowledge of machining processes and be familiar with the capabilities of the machine tool.

3-1 INTRODUCTION

NC part programming comprises the collection of all data required to produce the part, the calculation of a tool path along which the machine operations will be performed, and the arrangement of those given and calculated data in a standard format, which could be converted to an acceptable form for a particular MCU.

The necessary data for producing a part may be classified as follows:

1. Information taken directly from a drawing: dimensions, such as length, width, height, radius, etc.; segment shape: linear, circular, or parabolic; diameter of holes to be drilled. The tool path is calculated based on this information. *From Blueprint*
2. Machining parameters, which depend on surface quality, required tolerances, and type of workpiece and cutting tool: feeds, cutting speeds, and auxiliary functions such as turning on and off the coolant.
3. Data determined by the part programmer, such as the cutting direction and changing of tools. Part programmers establish the optimal sequence of operations which is required to produce the part. Therefore, they must be familiar with manufacturing processes and have detailed knowledge of the characteristics of the particular NC system.

4. Information depending on the particular NC system, such as acceleration and deceleration intervals or programming a two-spindle machine. When using computer languages for programming, those functions are performed by a postprocessor program.

All data is fed into the NC system in such a form that it can be read and processed by the MCU. Since the control medium for the NC system is standardized only when a perforated tape is used (EIA standards RS-273-A and RS-274-B) and since perforated tape is used today in most NC and CNC machines, only punched-tape programming will be discussed in this chapter.

There are two types of data processing techniques employed to produce the punched tape:

1. Manual (with the use of a calculator)
2. Computer-assisted preparation of tapes

Each of these techniques requires different stages in the preparation of the punched tape, as shown in Fig. 3-1.

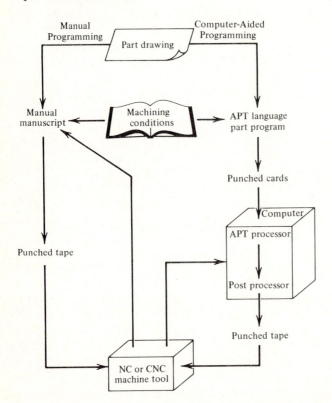

Figure 3-1 Comparison between manual and computer-aided part programming.

In manual programming the data required for machining a part is written in a standard format on a special manuscript. The manuscript is a planning chart or list of instructions which describes the operations necessary to produce the part. An example of a manuscript form for manual programming is shown in Fig. 3-2. Each horizontal

Machine: Drilling machine no. 2

Part name: Sample	Remarks:	Prepared by: J. Clark
	Use 3 tools;	Date: 4/25/82
Part no.: Fig. 3-3	center drill	
	8-mm-diameter drill	Checked by: Y. Koren
	20-mm-diameter drill	

N seq. no.	TAB or EB	Sign .	X increment	TAB or EB	Sign	Y increment	TAB or EB	M fun.	EB
000								RWS	EB
001	TAB	+	4500	TAB	-	1500	TAB	03	EB
002	TAB	+	5000	EB					
003	TAB	-	2500	TAB	-	4500	EB		
004	TAB	-	7000	TAB	+	6000	TAB	06	EB
005	TAB	+	4500	TAB	-	1500	TAB	03	EB
006	TAB	+	5000	EB					
007	TAB	-	9500	TAB	+	1500	TAB	06	EB
008	TAB	+	7000	TAB	-	6000	TAB	03	EB
009	TAB	-	7000	TAB	+	6000	TAB	30	EB
Check: ΣX = 0000					ΣY = 0000				

Handwritten annotations at right:
- { Center Drill (rows 001–003)
- } Tool Change (row 004)
- { 8mm Drill (rows 005–006)
- } Tool Change (row 007)
- } 20mm Drill (row 008)
- } Indicates end of Tape (row 009)

Figure 3-2 The NC programming manuscript for the part in Fig. 3-4.

19 Rows

line comprises a *block* of information, as was explained in Chap. 1. The manuscript is typed with a Teletype or Flexowriter, which produces the typed manuscript and the punched tape simultaneously. Each symbol (e.g., letter, number, algebraic sign) of the manuscript corresponds to one perforated row on the tape, which is referred to as a *character*. A group of characters representing an instruction or a complete piece of numerical data is referred to as a *word*.

Part programming of simple parts can be performed manually, with calculators to assist in trigonometric calculations. Therefore, manual programming is generally used for parts to be produced on point-to-point machines, in which the tool path calculations are straightforward and cutter radius compensations are not required. By contrast in milling, the MCU has to receive dimensional information about the cutter center. Figure 3-3 shows in solid lines the part contour, and in broken lines the contour which

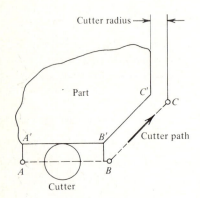

Figure 3-3 Cutter radius compensation is required to calculate the cutter center part.

defines the path of the cutter center after compensation for the radius. The cutter center path must be programmed. However, the calculation of this path is not simple, even for a linear segment such as $\overline{AB}$ in Fig. 3-3. When the part configuration is complicated, every motion of the cutter requires an enormous amount of mathematical calculations, and computing equipment much more sophisticated than a calculator becomes a necessity.

Computers are used as an aid in NC programming by applying special-purpose programming languages. The best known and most comprehensive one is the *A*utomatically *P*rogrammed *T*ools (APT) system, which facilitates the part programmer's work. It consists of a series of statements which are punched on cards and are used as the input to the computer. The APT language provides the same flexibility of expression to part programmers that other standard programming languages provide to computer programmers. Other known computer programming systems are ADAPT, SPLIT, EXAPT, AUTOSPOT, COMPACT II, and many others which have been developed by various companies.

3-2 MANUAL PROGRAMMING

3-2.1 Basic Concepts

Manual tape preparation is executed by part programmers. First they have to determine the machining parameters and the optimal sequence of operations to be performed. Based on this sequence, they calculate the tool path and write a manuscript, similar to the one shown in Fig. 3-2. Each line of the manuscript, which is referred to as a *block*, contains the required data for transferring the cutting tool from one point to the next one, including all machining instructions that should be executed either at the point or along the path between the points.

The EIA standard RS-273-A provides a line format for point-to-point and straight-cut NC machines. A typical line according to this standard, is as follows:

$$n102 \ g04 \ x-52000 \ y09100 \ f315 \ s717 \ t65432 \ m03 \ (EB) \qquad (3-1)$$

The letter and the number which follows it are referred to as a *word*. For example, $x-52000$ and m03 are words. The first letter of the word, e.g., x, is the *word address*. The word addresses are denoted as follows: n, sequential number; g, preparatory function; x and y, dimensional words; f, feedrate code; s, spindle speed code; t, tool code; m, miscellaneous function; and (EB) is the end-of-block character. The EB character is not printed, but only punched, and it is usually the carriage-return code, thus permitting a new line to begin immediately afterward. The EB indicates to the MCU that the present reading is completed and the axes of motion have to start up. When this motion is accomplished the next block is read.

The EIA standards RS-273-A and RS-274-B describe a *variable block format,* which is a combination of word address and tab sequential format and includes both tab and address characters. A practical NC system, however, is designed to use either tab or address characters, but not both.

In the *word address format* each word must be headed by the word address. The MCU uses the address letter to identify the word which follows it. In this type of format, words need not be arranged in any specific order within the block, since the letter identifies the corresponding word. However, since the standards prescribe a definite sequencing of words, the format recommended in (3-1) is used in practice for point-to-point programming.

In the *tab sequential format* a tab character (punched by depressing the tabulator key of the Flexowriter) is inserted between each two words in the block, and the address letter can be omitted. In this format, words must be arranged in a specific order, like the one suggested in (3-1) for point-to-point programming. When a word is not needed in a particular block, it may be omitted, but the corresponding tab character must be punched.

The block format of (3.1) includes all leading and trailing zeros in the dimension words (x and y). A system may be designed to permit elimination of trailing zeros (which follow significant digits) or ignoring leading zeros (which precede the significant digits). In the case where trailing zeros are ignored, the leading zeros must be punched and vice versa. The + (plus) sign in dimension words may be ignored in most systems, where no sign means that a positive coordinate, or positive increment, would be used.

A simple part for drilling has been drawn in Fig. 3-4 to illustrate the manual part

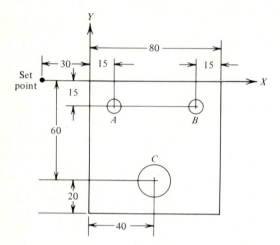

Figure 3-4 Part for drilling.

programming process. First a zero point (origin of coordinates) was selected, and a *set point* was established. This point is used for tool change and for loading new parts, and therefore it is not located above the part surface. In this part, the center of each hole must be drilled prior to the final drilling with the required diameter. The spindle speed and the drilling feed in this system are controlled from the operator console and are not programmed. The system operates as follows. Upon reaching the correct position on the *XY* plane, the drill advances toward the workpiece, drills the hole, and retracts. This retraction provides a restart signal, and the next block is read automatically. However, in cases where the spindle is instructed to stop, the tool does not advance toward the workpiece and the system halts.

The programming requires the use of several miscellaneous (m) functions which are listed below:

m03 Starts spindle rotation in clockwise direction.
m06 Indicates the necessity of a tool change. At the end of the block, the spindle is automatically stopped and a tool-change indicator, located on the operator console, lights up. The operator changes the tool and restarts the system by pushing a start button.
m30 Indicates end of tape and is used to reset the control. Resetting includes rewinding the tape to the rewind-stop (RWS) character, clearing the sequence number register, and stopping the spindle. This function is executed after the completion of all commands in the block.

RWS Rewind-stop code, usually placed as the first character on the tape. This code stops the tape reader when it is automatically rewinding while executing the m30 function.

The system applies the tab sequential format and incremental programming; its resolution is BLU = 0.01 mm. The dimensions in Fig. 3-4 are in millimeters, but the dimensional words in the program are expressed in BLUs. The corresponding part program is shown in Fig. 3-2.

To start producing parts the operator moves the tool manually to the set point, loads the tape into the tape reader, and starts the machine. The NC system executes the program block by block as follows:

001 The spindle starts to rotate and the drill moves from the set point to hole A. The hole is center-drilled automatically.

002 The drill moves in the X direction from hole A to hole B. Hole B is center-drilled. Note that the y dimension is not programmed in this block, since its incremental value is zero.

003 The tool moves from hole B to hole C; hole C is automatically center-drilled.

004 The drill returns to the set point, the spindle stops, and the tool-change indicator lights up. The drill is changed by the operator, and the machine is restarted by pressing an appropriate button. The next block is then read.

005 to 007 The tool moves to holes A and B where the drilling of holes of the required diameter is performed automatically. The 007 block returns the tool to the set point and calls for another tool change. The operator changes the tool and restarts the machine.

008 After positioning, hole C is drilled.

009 The tool returns to the set point, and then the tape is rewound to the RWS code. The control is again ready to begin a new part.

This example was very simple. In more complicated cases, where many holes with various diameters have to be drilled, the part programmer has to find the most economical sequence of operations in order to minimize the entire traveling distance. Note also that in the above system the spindle speed and drilling feed could be varied only from the operator console. In other systems the speed, feed, and other functions are controlled by the tape. This is done by programming a code number for each spindle speed, and feed, as will be explained below.

3-2.2 Tape Format

With the increasing use of NC systems which apply perforated tape as the input medium, it becomes necessary to issue standards for the tape format. Based upon this need, the EIA has prepared the standards RS-273-A (May 1967) and RS-274-B (May 1967, revised in February 1969) which are concerned with "Interchangeable Perforated Tape Variable Block Format for Positioning and Straight Cut (RS-273-A) and for Contouring and Contouring/Positioning (RS-274-B) Numerically Controlled

Machines." The functions, or words, of which a variable block format consists, are explained in this section in detail.

Each line of the manuscript, which was already introduced in the previous section, is equivalent to one block. A full block format contains the following words, which are permitted to appear only in the order specified below:

1. Sequence number word n consists of three digits, is first in the block, and is followed by the preparatory word.
2. Preparatory function word g consists of two digits, followed by the dimension words.
3. Dimension words—arranged in the order x, y, z, u, v, w, p, q, r, i, j, k, a, b, c, d, e—for systems containing multiple axes. In a three-linear-axes machine the dimension words are in the order x, y, z, i, j, k. The last dimension word is followed by the feed word.
4. Feed word f is expressed by four digits when using the "inverse-time" method, or as a three-digit coded number when using the magic-three code, and in most CNC systems. The feed function is followed by the spindle-speed word.
5. Spindle-speed word s is usually expressed as a three-digit coded number obtained by using the magic-three code. It is followed by the tool word.
6. Tool word t consists of a maximum of five digits, and is followed by the miscellaneous word.
7. Miscellaneous function word m consists of two digits and immediately precedes the EB character.

The EB character is used to indicate the end of each block, and therefore it may be used after any complete word. On most tape-punching typewriters the carriage-return key, when depressed, punches the EB character and simultaneously begins a new line on the manuscript sheet.

The end-of-record character is usually used as the RWS code and is placed as the first character on the tape. This code will stop the tape reader when it is automatically rewinding after the part has been completed. The / (slash) character provides a block-delete function when it precedes the sequence number word.

For systems using the word address format, the word address (the letter) must be the first character of each word. Complete words may be omitted where not required. When using the tab sequential format, the tab character must be the last character of each word, except the last one in the block; word addresses are omitted. Words, except for the tab character, may be omitted where not required. This shall be interpreted by the MCU as no change in the previous state of the system with respect to the omitted word. Certain commands, which by their nature are completely executed in the block, shall be repeated each time they are required.

In the following pages the various words will be discussed in greater detail.

Sequence number n. Each block of the tape has a sequential number, the sequence number word, which is the first in the block. Its major purpose is to identify specific machining operations through the block number, particularly when testing a part program. When the tape is read, each sequence number is displayed at the operator console while the block commands are performed, thus enabling the operator to compare the actual performance with the programmed commands.

Preparatory function g. The preparatory function prepares the MCU circuits to perform a specific mode of operation. Therefore the preparatory function precedes the dimension word. The preparatory functions are given and explained in Table 3-1.

Table 3-1 Preparatory functions

Code	Function	Explanation
g00	Point to point, positioning	Use with combination point-to-point/contouring systems for indicating positioning operation.
g01	Linear interpolation (normal dimensions)	A mode of contouring control used for generating a slope or straight cut, where the incremental dimensions are normal, i.e., input resolution is as specified.
g02	Circular interpolation arc CW (normal dimensions)	A mode of contouring control which produces an arc of a circle by the coordinated motion of two axes. The curvature of the path (clockwise = g02, or counterclockwise = g03) is determined when viewing the plane of motion in the negative direction of the perpendicular axis. The distances to the arc center (i, j, k) are "normal dimensions."
g03	Circular interpolation arc CCW (normal dimensions)	
g04	Dwell	A programmed (or established) time delay, during which there is no machine motion. Its duration is adjusted elsewhere, usually by the f word. In this case dimension words should be set at zero.
g05	Hold	Machine motion stopped until terminated by an operator or interlock action.
g06	Parabolic interpolation (normal dimensions)	A mode of contouring control which uses the information contained in successive blocks to produce a segment of a parabola.
g08	Acceleration	The feedrate (axes' velocity) increases smoothly (usually exponentially) to the programmed rate, which is noted later in the same block.
g09	Deceleration	The feedrate decreases (usually exponentially) to a fixed percent of the programmed feedrate in the deceleration block.
g10	Linear interpolation (long dimensions = LD)	Similar to g01, except that all dimensions are multiplied by 10. For example, a programmed dimension of 9874 will produce a travel of 98740 basic length-units. (Used only with incremental programming.)
g11	Linear interpolation (short dimensions = SD)	As g01, but dividing all dimensions by 10, e.g., 987 units for the example above.

Code	Function	Explanation
g13 g14 g15 g16 } Axis selection		Used to direct the control system to operate on a specific axis or axes, as in a system in which controls are not to operate simultaneously.
g17	*XY* Plane selection	Used to identify the plane for such functions as circular interpolation or cutter compensation.
g18	*ZX* Plane selection	
g19	*YZ* Plane selection	
g20	Circular interpolation arc CW (LD)	As g02 with long dimension distances.
g21	Circular interpolation arc CW (SD)	As g02 with short dimension distances.
g30	Circular interpolation arc CCW (LD)	As g03 with long dimension distances.
g31	Circular interpolation arc CCW (SD)	As g03 with short dimension distances.
g33	Thread cutting, constant lead	A mode selected for machines equipped for thread cutting.
g34	Thread cutting, increasing lead	As g33, but when a constantly increasing lead is required.
g35	Thread cutting, decreasing lead	As g33, but to designate a constantly decreasing lead.
g40	Cutter compensation—cancel	Command which will discontinue any cutter compensation.
g41	Cutter compensation—left	Displacement, normal to cutter path, when the cutter is on the left side of the work surface, looking in the direction of cutter motion.
g42	Cutter compensation—right	Compensation when cutter on right side of work surface.
g43 through g49	Cutter compensation if used; otherwise unassigned.	Compensation (g40–g49) is used to adjust for difference between actual and programmed cutter radii or diameters.
g60 through g79	Reserved for positioning only	Reserved for point-to-point systems.
g80	Fixed cycle cancel	Command which will discontinue only fixed cycle.
g81 through g89	Fixed cycles #1 through #9, respectively.	A preset series of operations which direct the machine to complete such action as drilling or boring.
g90	Absolute dimension programming	A control mode in which the data input is in the form of absolute dimensions. Used with combination absolute/incremental systems.
g91	Incremental dimension programming	A control mode in which the data input is in the form of incremental dimension.

Dimension words. These words may be classified as follows:

1. Distance dimension words, whose address characters are
    (*a*) x, y, z for primary motion dimension in the *X*, *Y*, and *Z* directions, respectively
    (*b*) u, v, w for secondary motion dimension parallel to *X*, *Y*, *Z*, respectively
    (*c*) p, q, r for tertiary motion dimension parallel to *X*, *Y*, *Z*, respectively
2. Circular dimension words and thread-cutting dimension words both use the same address characters: i, j, k, for distances to the arc center or thread leads parallel to *X*, *Y*, *Z*, respectively. More details are given below.
3. Angular dimension words, whose address characters are
    (*a*) a, b, c for angular dimension around the *X*, *Y*, *Z* axes, respectively
    (*b*) d, e for angular dimension around other axes

Distance and angular words are expressed either by incremental or absolute coordinates; circular words are always incremental. All dimension words contain digital data as follows:

1. Decimal points are not used; dimension words are programmed in BLUs.
2. All angular dimensions are expressed in decimal parts of a revolution (recommended) or in decimal degrees.
3. An algebraic sign (+ or −) precedes the first digit, except in the following cases:
    (*a*) Where angular dimension words are programmed with absolute dimensions.
    (*b*) Where in an absolute system only positive coordinates are used.
    (*c*) Usually only negative signs must be programmed.
    (*d*) Usually i, j, and k words are programmed without algebraic signs.
4. In some systems leading zeros need not be programmed; in others trailing zeros may be omitted.

Circular interpolation Circular arcs are machined only in the main planes. The preparatory functions are used to select the plane (g17, g18, g19) and the direction of the tool along the arc (g02, g03, g20, g21, g30, g31).

Four dimension words per block are required. Two distance dimension words are used to provide the distances to the end of the arc, and two circular dimension words give the distances to the arc center. Usually i, j, and k express distances parallel to the *X*, *Y*, and *Z* axes, respectively.

For example, the dimension words portion in the block corresponding to the circular arc *AB* shown in Fig. 3-5 (BLU = 0.001 in) will be either

$$\text{x-01250} \quad \text{y01750} \quad \text{i01250} \quad \text{j00500} \tag{3-2}$$

when programmed in absolute dimensions (coordinates), or

$$\text{x00750} \quad \text{y00750} \quad \text{i01250} \quad \text{j00500} \tag{3-3}$$

when programmed in incremental dimensions.

An arc must end in the same quadrant in which it starts. Arcs which lie in more than one quadrant require two or more blocks in the part program.

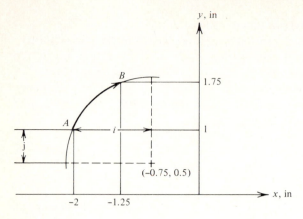

A = initial point
B = final point

Figure 3-5 Dimension words in a circular arc (BLU = 0.001 in).

Thread cutting The thread-cutting mode is selected by a preparatory function (g33, g34, g35). Up to four dimension words per block are required. The lead parallel to the X, Y, and Z axes is addressed by i, j, and k, respectively.

Feed word f. The f word is generally used in contouring or straight-cut systems. In point-to-point systems a constant maximum feedrate is generally used to move the tool from point to point. Feedrates of linear or circular motion are independent of spindle speed and are expressed by inches per minute or by millimeters per minute rather than by inches per revolution or millimeters per revolution, as in conventional machines.

Several methods for expressing the feedrate word are in use, and the most popular ones are explained below.

The inverse-time code The f word provides the MCU with a coded feedrate number (FRN), which is the reciprocal of the interpolation time in *minutes*. In linear motions it is calculated as follows. The velocity of the cutting tool along the path in length-units per minute (i.e., the feedrate) is multiplied by 10 and divided by the corresponding incremental distance measured in the same length-units:

$$\text{FRN} = 10 \times \frac{\text{feedrate along the path}}{\text{length of the path}} \tag{3-4}$$

The coded FRN is expressed by a four-digit number ranging from 0001 through 9999. This corresponds to an interpolation time of 10 min and 0.06 s, respectively. As an example, if it is required to move over a distance of 5 in at a feedrate of 10 in/min, the FRN would be

$$\text{FRN} = \frac{10 \times 10}{5} = 20$$

and the feed word is f0020. Note that the same result is achieved when the distance is 125 mm and the desired feedrate is 250 mm/min.

There are systems in which leading zeros are not necessary to be programmed; in this case the feed word will be f20.

The formula for calculating the FRN in circular motions is

$$\text{FRN} = 10 \times \frac{\text{feedrate around the arc}}{\text{radius of the arc}} \tag{3-5}$$

For example, the radius of the circular arc in Fig. 3-5 is 1.346 in. If the required feedrate along the part is 5in/min, Eq. (3-5) yields FRN = 37.

The NC part program provides the tool motions and not the configuration of the part. Therefore, in contouring systems, the dimension words should describe the path of the center of the cutting tool rather than the actual part contour. Similarly, the length of the path in Eq. (3-4) and the radius of the arc in Eq. (3-5) relate to the cutter center path. The feedrates in these equations are the velocities of the cutter center.

CNC feed word In most CNC systems the feedrate is programmed directly in millimeters per minute, or in inches per minute times 10, using a three-digit number. For example, a feedrate of 250 mm/min is programmed as f250, and a feedrate of 5.5 in/min is written as f55. Once the feed word enters the CNC system, it remains until replaced by another feed word.

The programmed feedrate in CNC systems should be the feedrate along the cutter center path. In linear motions, the feedrate of the cutting tool and the obtained part surface feedrate are equal. In circular motions, however, the required feedrate F_r should be multiplied by the ratio of the tool-path radius to the part contour radius:

$$F = \frac{R_p \pm R_t}{R_p} F_r \tag{3-6}$$

where F is the programmed feedrate, R_p is the part contour radius, and R_t is the tool radius. For cutting around the outside of a circle, the plus sign in Eq. (3-6) is used, and the tool feedrate is increased. For cutting around the inside of a circle, the minus sign is used, and the feedrate is decreased.

To illustrate the programming of the feedrate around circular contours assume that the shape given in Fig. 3-6 should be machined on a CNC machine at 6 in/min, using incremental programming and the word address format. The radii of the two arcs given in Fig. 3-6 are 1.5 in, the diameter of the cutter is 1 in, and the system resolution is BLU = 0.001 in. The two circular blocks in the program are as follows:

$$\text{g02 x2000 y-2000 i0 j2000 f80} \tag{3-7}$$

$$\text{g03 x1000 y-1000 i1000 j0 f40} \tag{3-8}$$

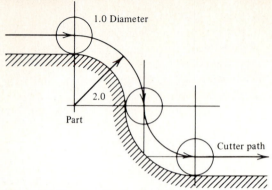

Figure 3-6 Compensation for the cutter radius in circular arcs.

The feed word in (3-7) was calculated as follows:

$$FRN = 10 \times 6 \times \frac{1.5 + 0.5}{1.5} = 80$$

The calculation of the feed word in (3-8) applies the minus sign in Eq. (3-6).

Spindle-speed word s. The spindle speed word is programmed in revolutions per minute and is expressed by a three-digit coded number, usually calculated by the magic-three code.

The magic-three code is calculated as follows:

1. The speed (in revolutions per minute) is rounded to two-digit accuracy; the second and third digits of the code are these two digits.
2. The speed is written in the form of the biggest decimal fraction multiplied by the power of 10; the first digit of the code has a value of 3 greater than the power of 10.

For example, 1645 rpm is rounded off to 1600 and written as 0.16×10^4, and the coded number will be 716. Similarly, a spindle speed of 742 rpm is given by the magic-three code as 674.

Tool word t. The tool word may consist of a maximum of five digits in a coded number. Each cutting tool has a different code number. The tool is automatically selected by the automatic tool changer when its code number is programmed in a block.

Several different methods for coding toolholders are used by the machine tool manufacturers. In these methods the manufacturer provides keys which correspond to the five-digit coded numbers; a key is provided for each toolholder. The key is comprised of five groups of lugs (or rings), where in each key some of these lugs have been removed to create the tool code. Each key is inserted in a cartridge aligned with the toolholder, in which the lugs push out metal springs which activate contacts, which produce a coded signal. When the programmed t number and the signal from the cartridge coincide, the appropriate toolholder is selected and a tool change is automatically performed.

Miscellaneous function m. The miscellaneous function words consist of two digits. This function pertains to auxiliary information which does not relate to dimensional movement of the machine, such as spindle command, coolant on and off, and other functions which are explained in Table 3-2. It should be noted that the m00, m01, m02, m06, and m30 are executed only after completion of all other commands in the block.

Table 3-2 Miscellaneous functions

Code	Function	Explanation
m00	Program stop	Stops spindle, coolant, and feed after completion of the block commands. It is necessary to push a button in order to continue the program.
m01	Optional (planned) stop	Similar to m00, but is performed only when the operator has previously pushed a button, otherwise the command is ignored.
m02	End of program	Indicates completion of the workpiece. It stops spindle, coolant, and feed after completion of all instructions in the block. May include rewinding of tape.
m03	Spindle CW	Starts spindle rotation in a clockwise direction.
m04	Spindle CCW	Starts spindle rotation in a counterclockwise direction.
m05	Spindle off	Stops spindle; coolant turned off.
m06	Tool change	Executes the change of a tool (tools) manually or automatically, not to include tool selection.
m07	Coolant no. 2 on	Turns a flood coolant on.
m08	Coolant no. 1 on	Turns a mist coolant on.
m09	Coolant off	Automatically shuts the coolant off.
m10	Clamp	Automatically clamps the machine slides, workpiece, fixture, spindle, etc. (as specified by the producer).
m11	Unclamp	Unclamping command.
m13 m14	Spindle CW & coolant on } Spindle CCW & coolant on }	Combines spindle rotation and coolant on in the same command.
m15 m16	Motion + } Motion − }	Rapid traverse or feedrate motion in either the plus or minus direction.
m19	Oriented spindle stop	Causes the spindle to stop at a predetermined angular position.
m30	End of tape	Similar to m02 except that it must include rewinding of tape to the rewind-stop character, thus ready for next workpiece.
m31	Interlock bypass	Temporarily circumvents normal interlock.
m32 through m35	Constant cutting speed	The control maintains a constant cutting speed by adjusting the rotation speed of the workpiece inversely proportional to the distance of the tool from the center of rotation. Normally used with turning.
m40 through m45	Gear changes if used; otherwise unassigned	

Unassigned: m12, m17, m18, m20 to m29, m36 to m39, m46 to m99.

3-2.3 Contour Programming—Example

To demonstrate manual tape preparation techniques, the part shown in Fig. 3-7 has been selected. Although the part is relatively simple, many calculations are required, partic-

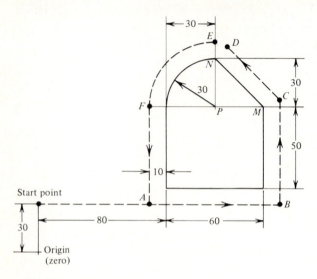

Figure 3-7 Part for milling.

ularly to determine the cutter center path. Machining is performed with an NC vertical milling machine equipped with a three-axis MCU which is capable of generating linear and circular motions. The characteristics of the NC system are as follows:

1. The tape format corresponds with the EIA standard RS-274-B; the tab sequential format is applied; and words are arranged in the following order: n, g, x, y, z, i, j, k, f, s, m.
2. The preparatory functions g00 through g04, and g17 through g19, are available.
3. Dimension words: maximum of five digits; the resolution is BLU = 0.01 mm; incremental programming; both leading and trailing zeros are programmed. When a dimension word is not programmed, it means that no change is required in the corresponding coordinate.
4. The feed word is a four-digit code, calculated with the inverse-time code. A constant machining feedrate of 500 mm/min is required. The rapid traverse feedrate is 2500 mm/min.
5. The spindle speed coding is the three-digit magic-three code. A constant spindle speed of 1740 rpm is required.
6. A tool word is not programmed, since only one cutter is in use.
7. The following miscellaneous functions are available: m03, m05, m06, m08, m09, and m30.

A cutter of 20 mm diameter was selected for this job. The cutter is initially located 40 mm above the table at the start point.

Figure 3-7 shows the part (dimensions in millimeters) and the path of the center of the cutter (the dashed line). Corner points are labeled for identification. The cutter center path should be offset from the part contour by the cutter radius, i.e., 10 mm. A reference zero (origin) is established for the coordinate system, which, in this case, is outside of the part itself.

A block-by-block description of the programming process is given below. Note that at the blocks shown below, the character / is used to symbolize the tab character. The first character on the tape is a rewind-stop code which is followed by an EB character. Next the motion blocks start.

From start point to point A. When moving from the start point toward point A, two blocks are programmed. At the first block, the system will accelerate to the traverse feedrate of 2500 mm/min. At the second block the machining feedrate 500 mm/min is programmed, and the motion decelerates before hitting the workpiece. At the end of these two blocks the center of the cutter is located at point A. A z dimension word must be programmed as well, to bring the cutter down to its appropriate place:

```
n     g    x        y       z                f       s      m
001 / 01 / 06000 / 00000 / - 04000 / / / / 0417 / 717 / 03
002 / 01 / 01000 /         /           / / / / 0500 /      / 08
```

The spindle starts at the first block, accelerates, and reaches 1700 rpm at the end of the block. The f words were calculated according to Eq. (3-4). The functions m03 and m08 start spindle rotation and coolant, respectively.

From point A to B. Motion proceeds in the X direction. The block is

```
n     g    x                 f
003 / 17 / 08000 / / / / / / 0063
```

The g17 code has to be placed in any block before the first circular arc.

From point B to C. The cutter moves in the $+Y$ direction to point C which is calculated with the aid of Fig. 3-8. A triangle $MB'C$ is formed, in which the angle $B'MC$ has a value of 22.5°. Therefore, the segment $\overline{CB'}$ is equal to

$$\overline{CB'} = \overline{MB'} \tan 22.5 = 10 \tan 22.5 = 4.14 \text{ mm}$$

and the segment $\overline{BC}$ is equal to

$$\overline{BC} = 10 + 50 + 4.14 = 64.14 \text{ mm}$$

The corresponding block is:

```
n     g    x        y                f
004 / 01 / 00000 / 06414 / / / / / 0078
```

From point C to D. From Fig. 3-8 we see that $x = y$ and that x may be calculated as follows:

$$x = \frac{CD}{\sqrt{2}} = 30 + \frac{4.14}{\sqrt{2}} = 32.93 \text{ mm}$$

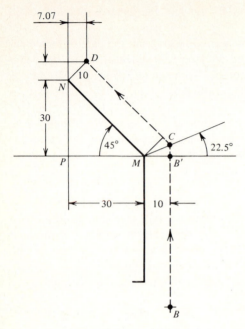

Figure 3-8 The path between points B and D in Fig. 3-7.

The corresponding block is

n   g   x   y   f
005 / 01 / − 03293 / 03293 / / / / / 0106

From point D to E. After passing point D and before starting the cutting of the circle EF, the cutter must move to point E in the circular arc ED, whose center is N. As illustrated by Fig. 3-9, the following equations have to be fulfilled:

$$i^2 + j^2 = R^2 \tag{3-9}$$

$$j + y = R \tag{3-10}$$

$$i = x = 0.707R \tag{3-11}$$

which, by substituting $R = 10$ mm, yield

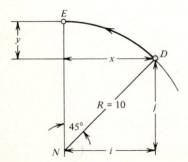

Figure 3-9 The circular arc between points D and E in Fig. 3-7.

$$x = i = j = 7.07 \text{ mm} \qquad y = 2.93 \text{ mm}$$

The corresponding block is

| n | g | x | y | i | j | f |

006 / 03 / $-$ 00707 / 00293 / / 00707 / 00707 / / 0167

Since in this block a surface cutting is not performed, any f word can be programmed. However, the present feedrate number was already computed for the next block. The velocity of the cutter center is $500 \times 40/30 = 667$ mm/min. According to Eq. (3-5)

$$\text{FRN} = 10 \times \frac{667}{40} = 167$$

From point E to F. The movement is along a circular arc of 90° whose radius is 40 mm:

| n | g | x | y | i | j | f |

007 / 03 / $-$ 04000 / $-$ 04000 / / 00000 / 04000 / / 0167

It should be noted that by using the f0167 code, the cutter center velocity is increased to 667 mm/min in order to maintain a surface feedrate of 500 mm/min.

From point F to A. This is a straight cut in the $-$ Y direction:

| n | g | x | y | | | f |

008 / 01 / 00000 / $-$ 06000 / / / / / 0083

From point A to start point. The rapid traverse feedrate is applied, and the block is

| n | g | x | y | z | f | m |

009 / 01 / $-$ 07000 / 00000 / 04000 / / / / 0357 / 30

The function m30 will turn off the spindle and coolant and rewind the tape to the beginning of the program.

The complete manuscript is shown in Fig. 3-10, in which T denotes the tab character. Note that the total sum of the x dimensions and that of the y dimensions is zero.

From this example, it becomes obvious that except for extremely simple parts, the assistance of a computer in handling the calculations in contouring programming is essential.

3-3 COMPUTER-AIDED PROGRAMMING

3-3.1 General Information

A computer support as an aid to part programming was not required during the early period of NC use. The parts to be machined were of two-dimensional configurations requiring simple mathematical calculations. With the increased use of NC systems and growth in complexity of parts to be machined, the part programmer was no longer able to calculate efficiently the required tool path, and the use of computers as an aid to part programming became a necessity. The computer allows the economical programming of the cutter center path of complex parts that could not be manually handled. Computers can perform the required mathematical calculations quickly and accurately, and the

	Part name	EXAMPLE
	Part number	1
	Sheet	1
	Remarks	

Manuscript contouring program

Prepared by Y. Koren   Date Nov. 26, 1931
Checked by _____   Date _____
Machine _____
Tape number _____

n	g	x	y	z	i	j	k	f	s	t	m	REMARKS
000	REWIND	STOP CODE										
001	01	+06000	00000	-04000				0417	717		03	START to
002	01	01000						0500			08	POINT A
003	17	08000						0063				A TO B
004	01	00000	06414					0078				B TO C
005	01	-03293	03293					0106				C TO D
006	03	-00707	00293		00707	00707		0167				D TO E
007	03	-04000	-04000		00000	04000		0167				E TO F
008	01	00000	-06000					0083				F TO A
009	01	-07000	00000	04000				0357			30	A TO ST.

Figure 3-10 The complete manuscript of the part shown in Fig. 3-7.

64

computation errors, so commonly appearing in manual calculations, are eliminated in computer-aided part programming.

The calculations required to produce the punched tape by the computer are performed by a programming software system contained in the computer. The programmer communicates with this system through the system language which is based on English words. The APT language is the most comprehensive and popular system for part programming. Actually APT denotes both the computer system (the program) and the language. To program a part with APT, the programmer first defines, in an English-like language, the geometric shape of the part itself, rather than the cutter center path. After defining the part, the part programmer describes the path that the cutting tool should follow, by writing appropriate instructions. In addition, the computer is given the required tolerances and tool parameters, such as diameter. The computer processes the information and performs calculations similar to the ones performed in manual part programming. It calculates the cutter center path, the feedrate codes, etc. The result of this computer processing is a file of data, which is translated by a postprocessor program to the coded instructions necessary to operate the NC system.

The postprocessor is a computer program which is needed for every MCU/machine-tool (MT) configuration. The postprocessor generates as output either the punched tape or information which can be directly applied to prepare the tape by using some standard peripheral equipment.

In addition to the APT system, there are many other programming systems that have been developed by computer companies or by machine tool manufacturers. A few of these systems are available on minicomputers, but most of them can be handled only on large general-purpose computers. The amount of assistance the computer gives in preparing NC tapes will depend on the computer programming system. Some of the more common ones are COMPACT II, ADAPT, EXAPT, AUTOSPOT, AUTO-PROMPT, and SPLIT.

3-3.2 Postprocessors

The aid of computers in the preparation of punched tapes may be divided into two stages:

1. Using a general processor, such as the APT programming system, which accepts the programmer statements and produces instructions to guide the tool with the appropriate machining parameters, such as spindle speed and feedrate.
2. An additional computer program, referred to as a postprocessor, which accepts as input the general processor output, and generates as output either the punched tape for a particular MCU/MT system, or information suitable for easily preparing the tape.

The postprocessor is needed, since when performing a computation by the general processor, any information related to a specific MCU/MT configuration is ignored. The postprocessor output must be able to produce a part within the specified tolerance, at the programmed feeds, and take into account the dynamic effects of the system, such as overshoots, and the geometric constraints of the machine tool. These are the reasons that an individual postprocessor is required for each type of MCU/MT configuration. The advantage of having two separate processors—the general system and the postprocessor—is the smaller effort which is required to adopt the postprocessor to a particular system than the effort which would be required to modify a computer programming system, such as APT.

Postprocessor elements. Each postprocessor includes five principal elements as illustrated in Fig. 3-11: input, motion, auxiliary, output, and control. A brief description of each element and their main features are given below.

Input The input element reads the programming system (such as APT) output. Reading may be performed directly or with the aid of an input medium such as punched cards or magnetic tape. The input element checks the input data for reliability and prints a list of the unprocessible information and auxiliary elements for subsequent processing.

Motion The motion element is the main portion of the postprocessor. It performs all instructions concerned with the tool movement. The motion element includes two functions usually denoted as the geometry and dynamic portions, or packages. The input to the postprocessor defines the parts in the righthand cartesian coordinate system. On the other hand, there are MCU/MT configurations in which other coordinate systems are used, as in the case of multiaxes machines. In such cases the geometry portion performs coordinate transformation into the required system. In order to ensure

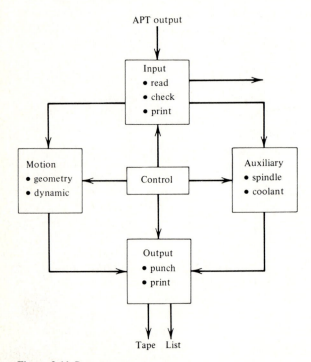

Figure 3-11 Postprocessor structure.

that the required tolerances are maintained, the geometry package checks the shape of the part in the new coordinate system. In the case where tolerances are out of range, new data points are generated along the path in order to maintain the specified tolerances. In addition, the geometry portion ensures that movement instructions to the MCU will not exceed the table size.

The dynamic package must prevent overshoots, undershoots, and other dynamic effects which are related to the NC system structure, in order to maintain the prescribed tolerances. This is done by modifying feedrates where necessary and establishing the distances for acceleration and deceleration.

Auxiliary The auxiliary element compares available preparatory and miscellaneous functions of a specific NC configuration with the required functions which are accepted from the input element. It determines whether each function is applicable to the MCU. In the case that the function is applicable, it is transferred to the output element in order to prepare a correct format output.

Output The output element receives data from the motion and auxiliary elements. This data is converted into a format appropriate to be accepted by the specific MCU. The output element generates either the punched tape or an output in another form that can be easily converted to a punched tape. In addition, it prints the list of the part program.

Control The control element generates the timing of the postprocessing, thus adapting all the elements and permitting program flow. It also controls the flow of data to the external output and the acceptance of new data for postprocessing.

3-4 APT PROGRAMMING

A great number of computer systems have been developed for NC programming. The APT system, however, is the most widespread and the most comprehensive one. The APT system is available on many computers and is widely used by many manufacturers of NC systems.

The first prototype of the APT system was developed, in 1956, by the Electronic System Laboratory of the Massachusetts Institute of Technology (MIT). Following this, the program was further developed by the cooperative efforts of 21 industrial companies sponsored by the Aerospace Industries Association (AIA) with assistance from MIT. As a result of these efforts, a system called APT II was produced in 1958, and a more effective system, the APT III, was distributed in 1961. The Illinois Institute of Technology Research Institute (IITRI) was selected to direct the future expansion of the program, and its capabilities are being continually expanded. The present APT language has a stock of approximately 300 words.

3-4.1 General Description

The APT program is a long series of instructions for a computer which specify the path that the tool must follow in order to produce a part. To communicate the tool path to the computer, one must provide the computer with geometric descriptions of the part surfaces. The APT language enables the programmer to do this and then to specify the way that the tool should move along these surfaces. The geometric description and the motion statements represent about 70 percent of the average program. An example of a geometric statement is

PT2 = POINT/3, 4	PT2 is the symbolic designation of a point whose *X* coordinate is 3 and whose *Y* coordinate is 4.

Examples of motion statements are:

GOTO/HOLE 2	Move the tool to the *X* and *Y* coordinates of a point called HOLE 2, which was defined elsewhere in the program.
GOLFT/L1, PAST, L2	Start moving to the left and then move the tool along a line called L1 until it passes a line called L2.

Note that PT2, L1, and HOLE 2 are identifying names. Identifying names are given to geometric expressions, such as points, lines, etc., and cannot be APT words. In addition to the geometric and motion statements there are other kinds of statements and features. One of the most useful statements is the CLPRNT. The CLPRNT is an instruction to the APT system to produce a printed list of all the cutter location coordinates that have been computed. The computation results are those of the APT system, before postprocessing.

Most APT statements are divided into two sections, major and minor, which are separated by a slash. The major section appears to the left of the slash and generally is one word, containing from one to six letters. The minor section, if required, appears to the right of the slash and contains modifiers to the major portion. For example, in the last statement GOLFT is the major section, and PAST is a modifier.

The APT language enables the definition and machining of three-dimensional (3-D) surfaces. However, since the intent of this text is to explain only principles of programming, it contains a description of statements which enable the machining of 2-D parts.

3-4.2 Geometric Expressions

A geometric expression defines a geometric shape or form. For each geometric form there are from 1 to 14 different methods of definition. APT contains definitions for 16 different geometric forms, where the most useful ones are POINT, LINE, PLANE, CIRCLE, CYLNDR, ELLIPS, HYPERB, CONE, and SPHERE. Several examples for definitions of the first four forms are presented below.

Points. In APT a point can be defined in 10 different ways; three of these are given below.

1. By coordinates
     POINT/X coordinate, Y coordinate, Z coordinate

Example
    PT1 = POINT/10.1, 5
Note: If no Z coordinate is given, it is assumed to be zero.
2. By the intersection of two lines
    POINT/INT OF, symbol for a line, symbol for a line
Example
    PT2 = POINT/INT OF, LIN 1, LIN 2
3. By a center of a circle
    POINT/CENTER, symbol for a circle
Example
    PT3 = POINT/CENTER, C1
Note: PT1, PT2, and PT3 are identifying words or names and may be used later in the program.

Lines. A line can be expressed in 13 different ways, three of which are defined below.

1. Through two points
    LINE/symbol for a point, symbol for a point
Example
    L1 = LINE/PT1, PT2
2. By a point and a tangent circle (Fig. 3-12)
    LINE/symbol for a point, $\frac{\text{LEFT}}{\text{RIGHT}}$, TANTO, symbol for a circle
Note: The modifiers LEFT or RIGHT are applied looking from the point toward the circle.
Examples
    L1 = LINE/P1, LEFT, TANTO, CIR1
    L2 = LINE/P1, RIGHT, TANTO, CIR1
3. Through a point and an angle with another line
    LINE/symbol for a point, ATANGL, angular value, symbol for a line
Example
    L1 = LINE/P1, ATANGL, 40, L2
In this example the angle between the given line (L2) and the newly defined line (L1) is 40°. The angular value is always specified in degrees and decimal fractions

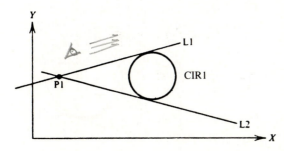

Figure 3-12 Line definition by a point and a tangent circle.

of a degree. The angle is positive if measured counterclockwise from L2 toward L1, and negative if measured clockwise. Therefore, the value -140 can be substituted for 40 in the above example.

Planes. Similar to points and lines, a plane can be defined in eight different ways; two of these are as follows:

1. By three points that are not on the same straight line
   PLANE/ symbol for a point, symbol for a point, symbol for a point
   Example
   PL1 = PLANE/ P1, P2, PT3
2. By a parallel plane and the perpendicular distance between the two planes

$$PLANE/ \ PARLEL, \text{symbol for a plane}, \ \begin{matrix} \text{XLARGE} \\ \text{XSMALL} \\ \text{YLARGE} \\ \text{YSMALL} \\ \text{ZLARGE} \\ \text{ZSMALL} \end{matrix} \ , \text{offset}$$

distance between the two planes
Example
PL2 = PLANE/ PARLEL, PL 1, ZSMALL, 5.1

In this example, PL2 is parallel to the given plane PL1, and is lower 5.1 units in Z. If the planes are not parallel to the main planes, two modifiers are appropriate, and either one of them could be used in the statement.

Circles. A circle can be expressed in 10 different ways; three of these are defined below.

1. By three points through which the circle is passing
   CIRCLE/ symbol for a point, symbol for a point, symbol for a point
   Example
   C1 = CIRCLE/PN2, (POINT/ 5.5, 7, 4.1), PNT1
2. By the center and a point on the circumference
   CIRCLE/CENTER, symbol for a circle center point, symbol for a point on the
      circumference
   Example
   C2 = CIRCLE/CENTER, (POINT/9, 7, 3), PT1
3. By the center and the radius.
   CIRCLE/CENTER, symbol for a circle center point, RADIUS, radius of circle
   Example
   C3 = CIRCLE/CENTER, PT1, RADIUS, 3

The CIRCLE/ statement defines in fact a circular cylinder perpendicular to the *XY* plane. The CYLNDR/ statement is used to define a cylindrical surface that could not be defined with a CIRCLE/ statement.

Example 3-1 Write in APT language the geometric expressions of the part shown in Fig. 3-13.

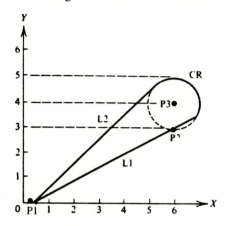

Figure 3-13 The part of Example 3-1.

SOLUTION

    P1 = POINT/0.5,0
    P2 = POINT/6.0,3.0
    P3 = POINT/6.0,4.0
    L1 = LINE/P1,P2
    CR = CIRCLE/CENTER, P3,P2
    L2 = LINE/P1, LEFT, TANTO, CR
    PL = PLANE/P1,P2,P3

3-4.3 Motion Statements

Once the required part has been defined with the geometric expressions, tool movement is specified using motion statements. Each motion statement will move the tool either to a new location or along a surface specified by the statement. Two groups of motion statements are available: for point-to-point and for contouring operations.

Point-to-point motion statements. Three motion statements exist for positioning the tool at a desired point, and their format is as follows:

FROM/symbol for a defined point	Indicates the initial position of the cutter center
GOTO/symbol for a defined point	Positions the tool center at a specified point
GODLTA/ΔX, ΔY, ΔZ	Positions the cutter in the specified increment from its current location

Note: (1) Instead of "symbol for a defined point," a statement (POINT/X, Y, Z) or X, Y, Z coordinates may be written. (2) FROM provides the initial location from which a motion is to start and is placed as the first motion statement in the program. The operator adjusts the machine to make the location of the tool coincide with this programmed location. (3) The GOTO/statement will move the tool along a path from the present location to the specified point. The GODLTA/ statement will move the tool the specified *incremental* distance from its present location. (4) In a drilling operation the GOTO/ statement is used to position the cutter above the required hole. A GODLTA/ statement is then used to plunge the cutter down into the workpiece. Another GODLTA/ statement is used to retract the tool. This sequence is repeated for each hole.

Example 3-2 Write a program to drill a hole, which is located at point (1, 1) from the set point (SETPT). The required depth is 0.5 in.

SOLUTION
    FROM/SETPT
    GOTO/1, 1, 0
    GODLTA/0, 0, − 0.5
    GODLTA/0, 0, 0.5
    GOTO/SETPT

Contouring motion statements. In APT programming, it is assumed that the part remains stationary and the tool moves. Three surfaces control the tool motion in contouring: the tool end moves on the *part surface*, the tool slides along the *drive surface*, and the motion continues until the tool encounters the *check surface*. The surfaces are shown in Fig. 3-14. However, before the tool can move along the controlling surfaces, it must be brought to them. This is executed by the *initial motion statement*, which has the following format:
    GO/cutter specifier, drive surface, cutter specifier, part surface, cutter specifier, check surface.
Three variations of cutter specifier, which are illustrated in Fig. 3-15, can be used:
    TO   ON   PAST

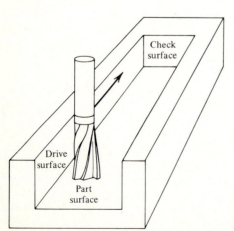

Figure 3-14 The surfaces which control the cutter in APT.

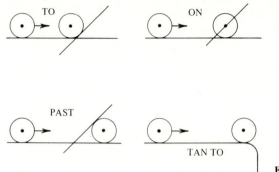

Figure **3-15** Cutter specifiers.

Example: GO/TO, CIRC1, ON, PL1, TO, LIN1

The drive surface of a GO/ statement will be the surface cut along in the next motion statement. The part surface is established for all the subsequent motion statements. The initial motion statement appears only once in a part program, and it brings the cutter from the set point to the workpiece. The actual cutting is controlled by another type of statement denoted as the *intermediate motion statements*.

Four variants of intermediate contouring motion statements may be used in APT. The most useful one has the following format:

motion word/ drive surface, cutter specifier, check surface

Example: GOLFT/DRS, TO, CKS

The drive surface is the surface cut along, and the check surface defines the end of the cutter motion.

Six different motion words, which are illustrated in Fig. 3-16, exist:

GOLFT GORGT GOFWD GOBACK GOUP GODOWN

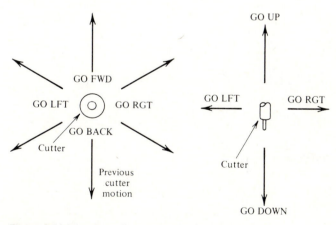

Figure 3-16 Directions of contouring motion words.

In intermediate motion statements, four different types of cutter specifiers, which are illustrated in Fig. 3-15, can be used:

TO   ON   PAST   TANTO

Note: (1) Motion words are programmed from the tool's viewpoint. (2) Each motion statement is dependent upon the preceding statement for establishing the direction of motion. (3) The check surface for the current motion is usually the drive surface of the next statement.

Example 3-3 Write in APT language the motion statements required to machine the part shown in Fig. 3-17. The part surface is called PLN.

SOLUTION

        GO/TO, L1, ON, PLN, ON, LL
        GORGT/L1, TO, L2
        GORGT/L2, TANTO, C1
        GOFWD/C1, TANTO, L3
        GOFWD/L3, PAST, L4
        GOLFT/L4, PAST, L5
        GOLFT/L5, PAST, L6
        GOLFT/L6, PAST, LL
        GOLFT/LL, PAST, L1
        GOTO/SP

Note the different use of the GO/TO and GOTO/ statements.

3-4.4 Additional APT Statements

Geometric expressions and motion statements represent about two-thirds of an average program. There are other kinds of statements which are required to accomplish an APT program.

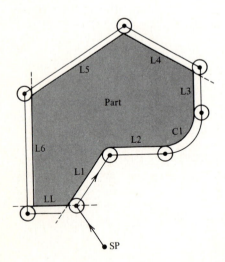

Figure 3-17 Part geometry for Example 3-3.

Postprocessor statements. One of the first statements of an APT program defines the postprocessor to be used by the APT. The structure of this statement is as follows:
MACHIN/ postprocessor name
Example: MACHIN/UN1
The postprocessor allows the transformation of a postprocessor control statement into an appropriate code of the control. Examples of postprocessor control statements are:

COOLNT/ON	Turn on the coolant (m08 code)
SPINDL/ON	Turn on the spindle (m03 code)
FEDRAT/25	Tool feedrate is 25 in/min
SPINDL/1250, CCLW	Spindle speed is 1250 rpm, in counterclockwise direction
TOOLNO/3572, 6	Tool number 3572, with 6 units in length
END	End of program (m02 code)

Tolerance and cutter specifications. All contouring motion commands are reduced to sequences of the straight-line motions required to approximate a given curve; the straight lines departing from the required contour surface by no more than a specified tolerance. The tolerance words are followed by an arithmetic parameter.
Examples

OUTTOL/.0005	Outer tolerance—affects overcutting; that is, the straight lines are tangent to the outside of the curve, departing by no more than 0.0005 units from the contour.
INTOL/.0001	Inner tolerance—affects undercutting.
TOLER/.005	Outer and inner tolerances are equal.

Cutter specifications are given by the statement CUTTER/ followed by up to seven arithmetic parameters, where the first one is the cutter diameter. In the APT programming example a simplified version of this statement is used:
CUTTER/10.0   Indicates a cutter diameter of 10 units.

Initial and Termination Statements. The first statement in an APT program begins with the word PARTNO. Any information may follow on the same line after PARTNO, for example
PARTNO PROGRAMMING EXAMPLE
The PARTNO statement identifies the name of the part.
The last statement in any APT program will be
FINI
It defines no successor and has the effect of terminating the program.

Conclusion. The APT vocabulary which is provided in this text is adequate to develop 2-D part programs for point-to-point and contouring systems. A complete vocabulary of APT permits the definition of the same geometry in a number of different ways, and thereby facilitates the task of the part programmer. The following example, which involves the development of a complete APT program, further illustrates the programming procedure.

3-4.5 An Example of APT Programming

In order to illustrate the programming procedure step by step, the part that was manually programmed in Sec. 3-2.3 is programmed using APT language. The dimensions of the part are illustrated in Fig. 3-7. The following rules may be helpful to the part programmer:

1. Spelling of APT words must be exactly as specified; note that only capital letters are used.
2. Generally, there is a punctuation mark (comma, equal sign, or slash) between every two words or numbers.
3. There will be no punctuation at the end of a statement.
4. A period is permissible only as a decimal point. A period at the end of a statement is an error.
5. Except for a few APT words (such as PARTNO), spaces can be inserted or omitted; e.g., it makes no difference if GO LFT or GOLFT is used.
6. It is recommended to use as identifying words (of geometric expressions) symbolic names that conjure associations with the geometric shape; e.g., use RITSID as the word to identify the right-side contour of a part.
7. An identifying word (name) cannot be an APT word.
8. Identifying words can be any combination of letters and numbers not exceeding six characters. The first character must be a letter.
9. Spaces can be inserted or omitted in the identifying words. Insertion or omission of spaces need not be the same throughout the same program; e.g., LIN2 can be used at the beginning and LIN 2 later.
10. Geometric expressions (shapes) must be defined and given an identifying word prior to using that identifying word in a statement. Therefore, it is recommended that definitions of the part geometry be placed before the motion statements.
11. Motion instructions are written from the cutter's viewpoint. That is, the cutter moves right or left as if the programmer were "riding" on the cutter and driving it along the desired path.
12. Use the sign $ (dollar) to indicate that a statement is continued on the next card.
13. A recommended structure of an APT program is as follows:

   PARTNO Part name and number
   MACHIN/ Postprocessor name
   Description and definition of the part geometry
   Cutter and tolerance specifications
   Machining conditions
   Motion statements
   Spindle and coolant off
   FINI

These general rules are applied to the specific example. Identifying names were assigned to the geometric elements and they are illustrated in Fig. 3-18. The direction

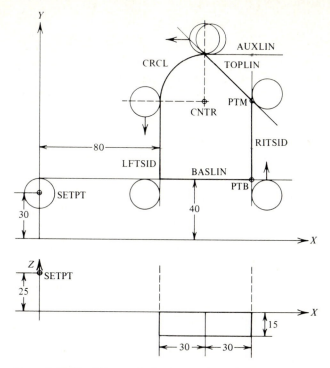

Figure 3-18 Identifying words for the part shown in Fig. 3-7.

of cutter motion is identical to that assigned in the manual programming example. The complete part program is given below.

The part program

PARTNO PROGRAMMING EXAMPLE	10
MACHIN/UN1	20
SETPT = POINT/0, 30, 25	30
CNTR = POINT/110, 90, *15*	40
CRCL = CIRCLE/CENTER, CNTR, RADIUS, 30	50
LFTSID = LINE/(POINT/80, 40), LEFT, TANTO, CRCL	60
PTB = POINT/140, 40	70
BASLIN = LINE/(POINT/80, 40), PTB	80
PTM = POINT/140, 90	90
RITSID = LINE/PTB, PTM	100
TOPLIN = LINE/PTM, (POINT/110, 120)	110
AUXLIN = LINE/(POINT/140, 120), RIGHT, TANTO, CRCL	120
XYPLN = PLANE/ CNTR, PTB, PTM	130
PSURF = PLANE/PARLEL, XYPLN, ZSMALL, 15	140

    $$ PART DESCRIPTION HAS NOW BEEN COMPLETED

CUTTER/20.0	150
TOLER/.005	160
SPINDL/1740, CLW	170
FEDRAT/2500	180
FROM/SETPT	190
GO / TO, BASLIN, TO, PSURF, TO, LFTSID	200
FEDRAT/500	210
GO FWD/ BASLIN, PAST, RITSID	220
GO LFT/ RITSID, PAST, TOPLIN	230
GO LFT/ TOPLIN, PAST, AUXLIN	240
GO LFT/ AUXLIN, TANTO, CRCL	250
GOFWD/ CRCL, TANTO, LFTSID	260
GOFWD/ LFTSID, PAST, BASLIN	270
FEDRAT/2500	280
GOTO/SETPT	290
STOP	300
FINI	310

Remarks

 10 Identifies the part name.

 20 Indicates the postprocessor name.

 30 SETPT is the startpoint of the tool. A part is replaced when the tool is at SETPT.

 40 This point is used in statement 50.

 50 Defines a circle called CRCL, with the point CNTR at its center, and a radius of 30 mm.

 60 Defines a line called LFTSID which passes through a point located at (80, 40), and is tangent to the left side of the circle CRCL when looking from the point toward the circle.

 70 Defines a point for subsequent geometric definitions.

 80 Defines a line called BASLIN that passes through the specified points.

 90 Defines a point which is used in the next statement.

100 Defines a line called RITSID on the right side of the part.

110 Defines a line called TOPLIN.

120 Defines a new line which is not included in the original drawing.

130 Defines a plane which coincides with the XY plane.

140 Defines a plane called PSURF which will be used later as the part surface, thus establishing the depth which is to be maintained while milling. PSURF is parallel to the XY plane and 15 mm below it.

150 A cutter with a diameter of 20 mm is to be used.

160 The cutter end will always be within 0.005 mm of the true mathematical surface along which it moves.

170 The spindle rotates in clockwise direction at a speed of 1740 rpm.

180 The desired feedrate is 2500 mm/min. The distance required for acceleration and deceleration (to hold overshoot within tolerances) is automatically computed when postprocessed.

190 Defines the initial location of the cutter.
200 Initial motion statement which moves the cutter to the surfaces named. Since PSURF is the part surface, the cutter will stay on it throughout the succeeding motions. BASLIN is the drive surface and has to be the surface cut along in the next motion statement.
210 the feedrate is reduced to 500 mm/min, which is the desired milling feedrate.
220 The cutter moves along BASLIN until passing RITSID.
230 The cutter moves to the left along the line RITSID until it is past TOPLIN.
240 Similar to 230.
250 The cutter moves along AUXLIN until it reaches the point where AUXLIN and CRCL are tangent.
260 Similar to 250.
270 Similar to 220.
280 Similar to 180.
290 The cutter moves to the point SETPT, which is out of the part.
300 The STOP statement stops the machine and turns off the spindle.
310 Indicates that the program is terminated.

3-5 OTHER PROGRAMMING SYSTEMS

In addition to the APT programming system, there are many other systems which operate more or less in the same manner. Some of the best known are summarized in Table 3-3, and a short description of several widespread systems is given below.

3-5.1 Description of COMPACT II

COMPACT II (*com*puter *p*rogram for *a*utomatically *c*ontrolling *t*ools) is a general-purpose processor developed by Manufacturing Data Systems, Inc., Ann Arbor, Michigan, a division of Schlumberger, Ltd. COMPACT II is used to write part programs for turning, milling, drilling, boring, machining center applications, flame cutting, and EDM. It is used to program parts for simple point-to-point drilling machines as well as complex three-, four-, and five-axis machines.

COMPACT II operates in an interactive environment on remote time-shared computers, in-house minicomputers, or in a batch environment on large in-house mainframe computers. The processor is machine-independent and operates with a machine-dependent set of subroutines, called a *link*, written for a specific machine tool and control unit. COMPACT II converts its language statements to manual NC codes in a single computer interaction. In contrast to languages requiring additional post-processing to complete the tape coding (e.g., APT), the link automatically performs the postprocessing as each statement is computer-processed. The link shortens the computational time and greatly facilitates the prompt discovery of detectable statement errors.

Table 3-3 NC computer programming systems

Program	Developed by	For
ADAPT	IBM (U.S. Air Force contract)	P, 3C
AUTOSPOT	IBM	P
CINTURN	Cincinnati Milacron	T
COMPACT II	MDSI, Ann Arbor, Mich.	P, $2\frac{1}{2}C$
EXAPT I	T.H. Aachen in Germany	P
EXAPT II	T.H. Aachen in Germany	T
EXAPT III	T.H. Aachen in Germany	3C
GENTURN	General Electric	T
MILTURN	Metaalinstitut in Netherlands	T
NEL 2PL	Ferranti	2P
NEL 2C	Ferranti	T
NEL 2CL	Ferranti	2C
PROMPT	Weber	C
SPLIT	White-Sundstrand Machine Tool	5P, 5C
UNIAPT	United Computing, Carson, Calif.	P, C

Note: C, contouring; *P,* positioning; *T,* turning operations; 2, 3, number of controlled axes.

COMPACT II statements. The language used in COMPACT II statements is English-like and similar to machine shop terminology. A COMPACT II program contains four types of statements: initialization, part description, tool selection, and tool motion. Initialization statements specify the machine tool link, the input-output mode, and the machine travel limits. Part description statements are used to define points, lines, circles, and planes. Either automatic or manual selection of tools may be specified in tool selection statements. Tool motion statements are used to describe both feedrate (CUT) and rapid traverse (MOVE) motion.

Processing involves informing the system of the machine-tool/control-unit to be used, passing each statement through a software link, and placing the resultant data in an output file. The output file data is punched into an NC part program tape, or the data is transferred directly to a CNC control. During processing, the user can interact with the computer to correct syntax errors, erroneous tool motions, and movements beyond the travel limits of the machine tool. The actual tool path can be plotted, along with the tool outline and part geometry. In this way, a correct part program can be created and checked in one computer session.

As in APT, the geometry definition and tool motion statements in COMPACT II are composed of major words, which describe what is to occur, and minor words, which describe how and where the occurrence is to take place. The major words for describing geometry include

DPTi—defines a point and stores its coordinates for later recall; i is the point's identification number.

DCIRi—defines a circle and stores the centerpoint coordinates and its radius for later recall; i is the circle's identification number.

The major words for describing tool motion include

MOVE—initiates rapid traverse motion to a specified location.
CUT—initiates feedrate motion to a specified location.
DRL—initiates a drilling cycle at a specified location.

A statement to define a point could be written as follows:

DPT1, 10XA, 6YA, 4ZA

where DPT1 defines a point and identifies it as point 1. 10XA is the X axis coordinate of the point, 10 units from absolute zero; 6YA is the Y axis coordinate of the point, 6 units from absolute zero; 4ZA is the Z axis coordinates of the point, 4 units from absolute zero.

Rapid traverse motion is programmed using a MOVE statement. To move the tool from the current point to point 1, you would simply program

MOVE, PT1

The COMPACT II software would generate the machine tool control codes to move the tool to the coordinate location of point 1. To produce a feedrate cut from point 1 (now the current tool position) to a perpendicular location tangent to the near side of line 1, you would program

CUT, TOLN1

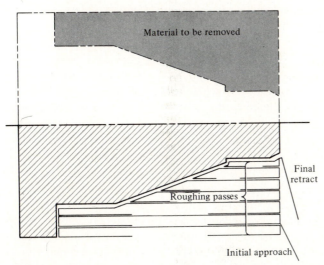

Figure 3-19 Material removal with COMPACT II's FasTurn instruction.

COMPACT II capabilities. COMPACT II provides machining capabilities for performing difficult specialized operations such as producing cams, performing machining operations on cylindrical parts, and producing machined lettering. COMPACT II also allows a program to be written for machining a complete group of similarly shaped parts called a *family of parts*.

Lathe features of the COMPACT II language include automatic cycles for threading and turning. The FasTurn is the COMPACT II material removal cycle for performing roughing and finishing operations on lathes. With FasTurn, the programmer defines boundaries representing the shape of the raw workpiece and finished part profile. The work area enclosed between the two shapes is the material to be removed by FasTurn (see Fig. 3-19). Complete material removal operations may be accomplished with a single COMPACT II input statement.

Similarly, milling, boring machine, flame cutter, and EDM features also include cycle modes which simplify the programming of parts with COMPACT II.

3-5.2 Additional Languages

ADAPT. The major disadvantage of the APT programming system is that it requires a large computer. The ADAPT (*ad*aption of APT) programming system was developed by IBM, under U.S. Air Force contract, in order to overcome this difficulty. Since ADAPT has a flexible modular structure, it is suitable for small- to medium-sized computers. By modular structure, we mean that subroutines may be omitted or added depending on computer size. The major advantage of the ADAPT system is that it uses a simplified version of the programming language used with the APT system. Usually an ADAPT program can be run on an APT system. However, since the APT system is much more powerful, it will not ordinarily run on ADAPT.

ADAPT is useful for programming of positioning and simple contouring parts in two or three dimensions, but it does not have the capability of programming a multiaxis contouring machine.

EXAPT. Some programming systems were developed in Europe, and the most widespread is the EXAPT (*ex*tended subset of APT) system which was developed in Germany at the Institutes of Technology in Aachen and Berlin. The EXAPT language is also compatible with APT, and it can use the same postprocessor available through APT. EXAPT has the capability of automatic selection of cutting speeds and feeds. This feature enlarges the required memory size of the computer and therefore is used only when large-scale computers are utilized. EXAPT has three variations: EXAPT I, for point-to-point machines; EXAPT II, for turning operations; EXAPT III, a 3-D contouring program.

AUTOSPOT. The AUTOSPOT (*auto*matic *s*ystem for *po*sitioning *t*ools) was developed by IBM and first introduced in 1962. It is one of the most popular programs available for positioning tasks such as drilling. AUTOSPOT permits programming in three dimensions with English-like input words and is used by many NC manufacturers.

BIBLIOGRAPHY

1. Brown, S. A., C. E. Drayton, and B. Mittman: A Description of the APT Language, *Commun. ACM*, vol. 6, no. 11, November 1963.
2. Childs, J. J.: "Principles of Numerical Control," Industrial Press, New York, 1965.
3. Conaway, J. O., D. Bringman, and R. C. Deane: "Handbook for SLO-SYN Numerical Control," The Superior Electric Company, Bristol, Connecticut, 1969.
4. EIA STANDARD RS-273-A, May 1967; STANDARD RS-274-B, May 1967; revised February 1969.
5. ITT Research Institute: "APT Part Programming," McGraw-Hill Book Company, New York, 1967.
6. Kelley, R. S.: The Production Man's Guide to APT-ADAPT, *Am. Mach.*, vol. 97, June 22, 1969.
7. Leslie, W. H. P.: Numerical Control Programming Languages, *Proc. PROCLAMAT-Congr.*, Rome, 1969, North-Holland Publishing Co., Amsterdam, 1970.
8. McWaters, J. F.: "NC User's Guide to the International Computer Programs," McGraw-Hill Publishing Company, London, 1968.
9. Stute, G.: "EXAPT, Möglichkeiten und Anwendung der automatisierten Programmierung für NC Maschinen," Carl Hanser, Munich, 1969.
10. Thomas, L. J.: "N/C Handbook," 3d ed., BENDIX Industrial Controls Division, Detroit, Michigan, January 1971.
11. Weil, R.: IFAPT a Unified System of Modular Design of NC-Languages, *IFIP-IFAC-PROCLAMAT-Congr.*, Rome, 1969, North-Holland Publishing Co., Amsterdam, 1970.

PROBLEMS

3-1 The dimensions of the part in Fig. 3-3 are $A'B' = 50$ mm, $B'C' = 40$ mm, and the angle $A'B'C' = 135°$. The required feedrate is 100 mm/min, and the inverse-time method is applied. Use the word address format (available: n, g, x, y, and f) to write a manual NC program of the lines corresponding to the segments $A'B'$ and $B'C'$. Assume that the radius of the cutter is negligible, the resolution is BLU = 0.01 mm, and incremental programming is used.

3-2 The radius of the cutter in Prob. 3-1 is 10 mm. Calculate the cutter center path, and write the manuscript lines which correspond to segments AB and BC.

3-3 Draw the part which was turned on a lathe using the following part program (BLU = 0.0001 in; only g and dimension words are given):

   g01   z-7500 x2500
   g01   z-10000
   g03   z-2500 x1250 k0 i3125
   g01   z-5000
   g01   x 1250
   g01 z-5000

The largest diameter of this part is 1.0 in; the g03 block is a counterclockwise circular cut, when looking in the positive Y direction.

3-4 The part given in Fig. 3-20 (dimensions in inches) is machined on a lathe in one turning cut at a constant feedrate of 4 in/min. Use the word address format, incremental programming, and the inverse-time code to write a NC manuscript. The system resolution is BLU = 0.0001 in; spindle speed and tool words are not programmed.

3-5 Write a NC part program, calculated manually, for milling the parts shown in Fig. 3-21 (dimensions in millimeters). Use tab sequential format, incremental programming, and BLU = 0.01 mm. Leading zeros are not programmed. Machining parameters: feedrate 120 mm/min (inverse-time code), and spindle speed, 1440 rpm (magic-three code). Cutter diameter is negligible. Note that the set point is at (20, 0) and the tool must travel back to this point at the end of the cut.

3-6 The part shown in Fig. 3-21a is machined on a CNC milling machine with a 10-mm-diameter cutter. Write the corresponding manual part program. Use the cutting conditions given in Prob. 3-5.

3-7 The part shown in Fig. 3-21b is machined on a CNC milling machine with a 5-mm-diameter cutter and at the cutting conditions specified in Prob. 3-5. Write the corresponding manual part program. Use Eq. (3-6) to calculate the feedrate.

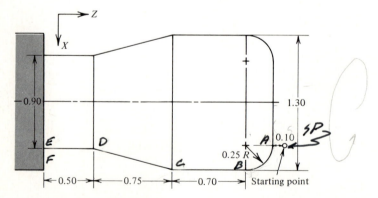

Figure 3-20 The part for Prob. 3-4.

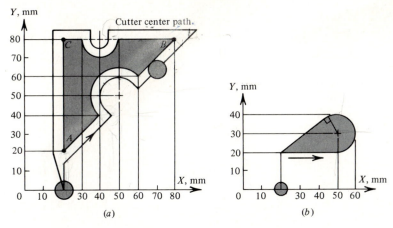

Figure 3-21 The parts for Probs. 3-5, 3-6, 3-7, 3-10, and 3-11.

3-8 The center of the circle C1 in Fig. 3-17 is the point *PC*, and its radius is 1.5 in. The intersection points between lines L1 and L2 is P1 and the angle between them is 45°. Points P1 and PC are defined. Write the geometric expression statements of C1, L2 and L1.

3-9 Write the tool motion statements for the part given in Fig. 3-13. Assume that the cutter is initially located at the origin.

3-10 Write an APT part program for the part in Fig. 3-21*b*; include the data given in Prob. 3-7. Use the following symbols: CRC for the circle, BASE for the baseline, and LL for the other line. The postprocessor is UOFM.

3-11 Write a part program in APT for the part shown in Fig. 3-21*a*; include the data given in Prob. 3-6. Use the following identifying names (symbols): line *AB* = LRT; line *BC* = LTOP; line *AC* = LFT; top circle = CT; other circle = CR. The postprocessor name is UOFM.

FOUR

SYSTEM DEVICES

This chapter discusses several components of NC, CNC, and robot systems. They can be divided into two groups: electromechanical devices and digital circuits. Every system includes a drive which converts the electrical command signals to mechanical motions. Closed-loop systems use feedback devices, which measure the axial motions and provide corresponding electrical signals. The MCU in NC and the computer in CNC and robots employ digital circuits of which counters and decoders are discussed. Digital-to-analog converters are also used in NC, CNC, and robot systems. They convert digital data to its continuous voltage counterpart, which, in turn, is used as the control signal to the axial drive.

4-1 DRIVES

Drives for NC and robot systems are either hydraulic actuators, dc motors, or stepping motors. The type selected is determined by the power requirements of the machine tool, the power sources available, and the desired dynamic characteristics.

Stepping motors are limited in power and available torque, and thus suitable only for small machine tools. DC motors provide excellent speed regulation, high torque, and high efficiency, and therefore they are ideally suited for control applications. DC motors can be designed to meet a wide range of power requirements and are utilized in most small- to medium-sized robots and NC machines. Hydraulic systems can range in size up to hundreds of horsepowers. They are well suited for large robots and NC machine tools where power requirements are high. The cost of a hydraulic drive is not proportional to the power required, and, thus, they are expensive for small- to medium-sized robots and NC machines.

4-1.1 Hydraulic Systems

Hydraulic systems are used extensively for driving high-power machine tools and industrial robots, since they can deliver large power while being relatively small in size. They can develop much higher maximum angular acceleration than dc motors of the same peak power. They have small time constants, and this results in smooth operation of the machine tool slides and the robot axes.

Hydraulic systems, however, present some problems in terms of maintenance and leakage of oil from the transmission lines and the system components. The oil must be kept clean and protected against contamination. Other undesirable features are the dynamic lags caused by the transmission lines and viscosity variations with oil temperature.

As shown in Fig. 4-1, hydraulic systems are generally comprised of the following components:

1. A hydraulic power supply
2. A servovalve for each axis of motion
3. A sump
4. A hydraulic motor for each axis of motion

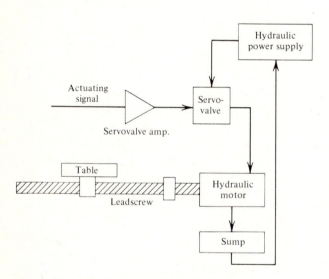

Figure 4-1 A general structure of a hydraulic system.

Hydraulic power supply. The hydraulic power supply is a source of high-pressure oil for the hydraulic motor and the servovalve. The main components of the hydraulic power supply are shown in Fig. 4-2 and are as follows:

1. A pump P for supplying the high-pressure oil. The frequently used types are the gear pump, and radial or axial displacement pumps.

2. An electric motor M, usually a three-phase induction motor, for driving the pump.
3. A fine filter for protecting the servosystem from any dirt or chips.
4. A coarse filter, located at the input of the pump, for protecting the latter against contamination that has entered the oil supply.
5. A check valve for eliminating a reverse flow from the accumulator into the pump.
6. A pressure-regulating valve for controlling the supply pressure to the servosystem.
7. An accumulator for storing hydraulic energy and for smoothing the pulsating flow. Accumulators can provide a large amount of energy over a short interval of time and are used where the load is characterized by an average demand which is far below the required peak. The accumulator supplies the peak requirements and is subsequently recharged by the pump. Another function of the accumulator is to smooth the pulsations caused by the pump, and the variations caused by the sudden motions of the valve. The accumulator functions like a capacitor in an electric circuit.

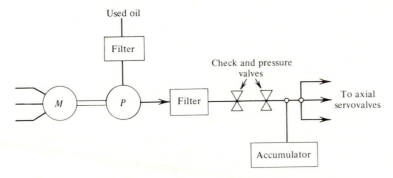

Figure 4-2 Hydraulic power supply.

Servovalve. The electro-hydraulic servovalve controls the flow of the high-pressure oil to the hydraulic motor. The servovalve receives a voltage-actuating signal and uses it to drive a solenoid device, which moves the valve spool. The magnitude of the input voltage V defines the flow rate of oil q through the valve. Assuming that the oil pressure drop across the valve is constant, the following relationship is valid:

$$q = K_v V \qquad (4\text{-}1)$$

where K_v is the valve constant. The flow rate of oil through the valve is proportional to the velocity of the hydraulic motor. The time constant of a servovalve, in a high-power system, is on the order of 5 ms, and is usually negligible compared with the other lags in the system.

The sump. The used oil is returned to a sump, or tank, through a special return line. The oil is fed back to the hydraulic power supply and forms the source of fluid for the latter.

The hydraulic motor. The hydraulic motor, or actuator, is either a hydraulic cylinder for linear motion or a rotary-type motor for angular motion.

The hydraulic cylinder, due to the large quantity of high-pressure oil which it contains, is limited to a relatively small motion. The rotary hydraulic motor is usually used in larger power servosystems. It operates at higher speeds and is geared down to the leadscrew which drives the table.

The simplified equation commonly used for determining the steady-state speed of hydraulic motors is

$$v = Kq \tag{4-2}$$

where q is the flow rate of oil through the valve (volume per time), v is the motor speed, and K is a constant.

4-1.2 Direct-Current Motors

Direct-current (dc) motors allow precise control of the speed over a wide operating range by manipulation of the voltage applied to the motor. They are ideally suited for driving the axes of small- to medium-sized NC machines and robots. DC motors are also used to drive the spindle in lathes and milling machines, when a continuous control of the spindle speed is desired.

The dc motor is actually a dc machine, which can function either as a motor or as a generator. The principle of operation of a dc machine is based on a rotation of an armature winding within a magnetic field. The armature is the rotating member, or *rotor,* and the field winding is the stationary member, or *stator.* The armature winding is connected to a commutator, which is a cylinder of insulated copper segments mounted on the rotor shaft. Stationary carbon brushes, which are connected to the machine terminals, are held against the commutator surface and enable the transfer of a dc current to the rotating winding.

For the case in which the dc machine serves as a generator, the rotor is turned at a constant speed by an external device connected to the shaft. The generator receives the mechanical energy from the shaft and converts it into electrical energy, which is provided through the armature output terminals. By contrast, in a motor, electrical energy is supplied to the armature from an external dc source, and the motor converts it to mechanical energy.

Two equations are required to define the behavior of a dc machine: the torque and the voltage equations. The torque equation relates the torque to the armature current:

$$T = K_f \phi I \tag{4-3}$$

and the voltage equation relates the induced voltage in the armature winding to the rotational speed:

$$E = K_f \phi \omega \tag{4-4}$$

where T = magnetic torque, N·m
ϕ = flux per pole, Wb
I = current in armature circuit, A

E = induced voltage (emf), V

ω = angular velocity, rad/s

K_f = constant determined by design of winding

For a motor, an input voltage V is supplied to the armature, and the corresponding voltage equation becomes

$$V - IR = K_f\,\phi\omega \qquad (4\text{-}5)$$

where R is the resistance of the armature circuit and IR is the voltage drop across this resistance. The armature inductance is negligible in Eq. (4-5).

Multiplying Eqs. (4-3) and (4-5) yields the power equation

$$P = \omega T = VI - I^2R \qquad (4\text{-}6)$$

where P is the mechanical output power, VI is the electrical input power, and I^2R is the electrical power loss.

DC motors are classified as separately excited, shunt-, series-, and compound-connected, according to the method of field connection. A schematic drawing of a

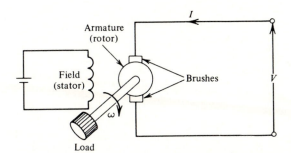

Figure 4-3 Schematic diagram of a separately excited dc motor.

separately excited motor with a load is shown in Fig. 4-3. The separately excited dc motor with constant field excitation is well suited for control applications, since it provides smooth control of speed over a wide range. The motor field is excited by a separate constant dc voltage supply, and consequently the flux ϕ becomes constant as well; thus Eqs. (4-3) and (4-5) are written as

$$T = K_t\,I \qquad (4\text{-}7)$$

$$V - IR = K_v\omega \qquad (4\text{-}8)$$

The parameters K_t and K_v are referred to as the torque and voltage constants. In SI units the torque constant in newton-meters per ampere equals the voltage constant in volt-seconds per radian. If English units are applied, the values of K_t and K_v become different. Many practical designs use technical meter-kilogram-second (mks) units, in which the speed is given in revolutions per minute and the torque in kilogram-meters. As a consequence K_t is given in kilogram-meters per ampere, and K_v in volts per revolution per minute. The ratio between these two constants is

$$\frac{K_v}{K_t} = \frac{2\pi g}{60} = 1.025$$

and in practice they are treated as equal constants.

Modern dc motors use a permanent-magnet (PM) field, rather than an externally excited field. Both types are referred to as dc servomotors and are characterized by Eqs. (4-7) and (4-8). The PM saves the field voltage source and results in higher efficiency and fewer thermal problems.

The dc servomotor drives a mechanical load consisting of dynamic and static components:

$$T = J\frac{d\omega}{dt} + T_s \tag{4-9}$$

where J is the combined moment of inertia of the motor and load, and T_s is the static load due to friction and cutting forces in NC systems.

Elimination of I and T from Eqs. (4-7) through (4-9), and rearrangement of the terms so as to separate the independent variables, gives the speed equation

$$\tau_m \frac{d\omega}{dt} + \omega = \frac{1}{K_v}V - \frac{R}{K_t K_v}T_s \tag{4-10}$$

where τ_m is the mechanical time constant of the loaded motor and is defined by

$$\tau_m = \frac{JR}{K_t K_v} \tag{4-11}$$

The Laplace transform of Eq. (4-10) is

$$\omega(s) = \frac{K_m V(s) - (RK_m/K_t)T_s(s)}{1 + s\tau_m} \tag{4-12}$$

where K_m is the gain of the motor and is defined by $K_m = 1/K_v$. The solution of Eq. (4-12) in the time domain depends on the applied voltage and load torque. For example, assuming that the motor is initially at rest, $T_s = 0$, and a step voltage of V volts is applied at the armature terminals, the solution is

$$\omega(t) = K_m V(1 - e^{-t/\tau_m}) \tag{4-13}$$

Thus the motor response is described by a steady-state speed $K_m V$ and a decaying exponential with a time constant τ_m given by Eq. (4-11).

Example 4-1 The voltage and torque constants of a PM servomotor are $K_v = 0.824$ V·s/rad, and $K_t = 7.29$ lb·in/A. The armature resistance is 0.41Ω and the armature inertia is 0.19 lb·in·s².

(a) Show that the voltage and torque constants are equal in SI units.

(b) Calculate the mechanical time constant of the motor.

(c) Calculate the steady-state speed for an 85-V input at no-load and full-load (120 lb·in) conditions; draw the corresponding speed versus time diagrams.

SOLUTION
(a) The torque constant is converted to SI units as follows:

$$K_t = 7.29 \times 0.0254 \times \frac{9.81}{2.205} = 0.824 \frac{\text{N·m}}{\text{A}}$$

and is equal to the voltage constant. Note that $g = 9.81$ m/s^2.
(b) From Eq. (4-11)

$$\tau_m = \frac{0.19 \times 0.41}{0.824 \times 7.29} = 13 \text{ ms}$$

(c) At no-load conditions

$$\omega = K_m V = \frac{85}{0.824} = 103 \text{ rad/s}$$

or $$\omega = 103 \times \frac{60}{2\pi} = 985 \text{ rpm}$$

For a loaded motor the change in speed, from Eq. (4-10), is

$$\Delta\omega = -\frac{RT_s}{K_t K_v} = -\frac{0.41 \times 120}{7.29 \times 0.824} = -8.2 \text{ rad/s} \qquad \text{or} -78 \text{ rpm}$$

The motor speed is reduced to 907 rpm. The corresponding response is given in Fig. 4-4.

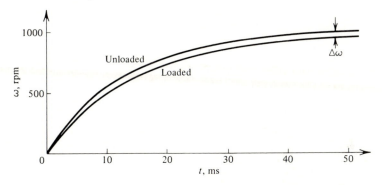

Figure 4-4 Time response of loaded and unloaded motor in Example 4-1.

Gearing. In many NC systems the leadscrew is driven through a gear box. The gear ratio K_g is defined as the ratio between the speed of the leadscrew ω_ℓ to the speed of the motor:

$$K_g = \frac{\omega_\ell}{\omega} \qquad (4\text{-}14)$$

In order to calculate the time constant by Eq. (4-11), the inertia of the leadscrew should be referred to the motor shaft. Consequently, the inertia J in Eq. (4-11) is

$$J = J_r + K_g^2 J_\ell \qquad (4\text{-}15)$$

where J_r is the inertia of the rotor and J_ℓ is the inertia of the leadscrew and load. Note that load torques should also be referred to the motor shaft:

$$T_s = K_g T_\ell \tag{4-16}$$

Where T_ℓ is the load torque at the leadscrew.

Example 4-2 The motor in Example 4-1 is coupled to the leadscrew of a PTP machine through a gear ratio of 2:1. The machine table weighs $W = 1000$ kg and the leadscrew pitch is $p = 10$ mm. If the static torque and the leadscrew inertia are negligible, calculate the torque required to accelerate the motor during $t = 0.15$ s, to a speed of $V = 150$ mm/s. Assume a constant acceleration rate.

SOLUTION The equivalent moment of inertia of the table is

$$J_t = \frac{Wp^2}{(2\pi)^2} \times 10^{-6} \text{ kg·m}^2 \tag{4-17}$$

Where the constant 10^{-6} converts square millimeters to square meters. Equation (4-17) yields $J_t = 2.53 \times 10^{-3}$ kg·m². The motor inertia converted to SI units is

$$J_r = \frac{0.19 \times 0.0254 \times 9.81}{2.205} = 0.19 \times 0.113 = 2.15 \times 10^{-2} \text{ kg·m}^2$$

From Eq. (4-15)

$$J = 2.15 \times 10^{-2} + \frac{2.53 \times 10^{-3}}{4} = 2.21 \times 10^{-2} \text{ kg·m}^2$$

The angular velocity of the motor is

$$\omega = \frac{2\pi V}{p} \text{ rad/s} \tag{4-18}$$

which yields $\omega = 94$ rad/s.

The torque required to accelerate the system is

$$T = \frac{J\omega}{\Delta t} = 13.8 \text{ N·m} \quad \text{or} \quad 122 \text{ lb·in}$$

Note: Multiply newton-meters by 8.85 to obtain pound-inches.

4-1.3 Stepping Motors

The stepping motor (SM) is an incremental digital drive. It translates an input pulse sequence into a proportional angular movement and rotates one angular increment, or a *step*, for each input pulse. The shaft position is determined by the number of pulses, and its speed is proportional to the pulse frequency. The shaft speed in steps per second is equal to the input frequency in pulses per second (pps).

Stepping motors can be used as the drive devices in open-loop NC systems. Since no feedback element is required, the system is cheaper than its closed-loop counterpart. However, the accuracy of the system depends upon the motor's ability to step through the exact number of pulses sent to its input. In addition, stepping motors are limited in torque and tend to be noisy and, therefore, are seldom used in practice.

To obtain optimal stepping motor performance, an electronic switch, or *translator,* is required as part of the drive unit. The drive unit contains a steering circuit and a power amplifier. The steering unit translates the incoming pulses into the correct switching sequence required to step the motor. The steered pulses are converted to power pulses, with appropriate rise time, duration, and amplitude for driving the motor windings. To reverse the motion, an additional input is provided. A 0 logic level at the latter causes a clockwise rotation and a 1 logic level—a counterclockwise rotation. The torque versus stepping rate characteristic, which is shown in Fig. 4-5, has a major

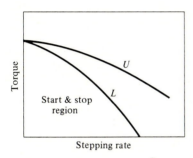

Figure 4-5 Torque versus stepping rate characteristic of a stepping motor.

significance in the design stage. The torque always decreases with an increase in the stepping rate. The exact shape of the curves depends on the driving unit and the stepping motor itself. The characteristic comprises two curves: the lower L and the upper U, sometimes denoted as the pull-in and pull-out curves, respectively. The upper curve shows the maximum running speed, and the lower curve shows the maximum allowable starting speed of the motor. In the allowable starting region, which lies between the lower curve and the axis, the motor can start, stop, and be reversed without missing a step. Once the motor has started, the stepping rate can be gradually increased toward the upper curve. Starting, stopping, or reversing the direction of rotation are not permitted in the region between the lower and the upper curves. Acceleration and deceleration should be performed gradually within this region. For example, to stop accurately, the pulse rate must first be gradually reduced to the rate prescribed by L, and then stopped.

Example 4-3 The torque speed characteristics of a stepping motor drive system are as follows:

$$L: \quad T = 200 - 0.050V$$
$$U: \quad T = 200 - 0.025V$$

where T is in ounce-inches and V is in steps per second. The motor is loaded by a constant torque of 70 oz·in.

(a) Calculate the maximum allowable starting speed and running speed of the motor.

(b) The motor runs at the starting speed during the first 13 pulses. What is the starting time period?

SOLUTION

(a) The maximum allowable starting speed is determined by L:

$$70 = 200 - 0.05V$$

which gives $V = 2600$ steps per second.

The running speed is determined by U, which yields $V = 5200$ steps per second.

(b) During the starting period the frequency is 2600 pps; therefore

$$t = \frac{13}{2600} = 5 \text{ ms}$$

In order to design a system with a maximum allowable speed, the stepping motor must be loaded with a constant torque. Stepping motors cannot be applied to contouring machines, since during cutting the load varies with the machining conditions. They are suited for driving small point-to-point machines in which the load torque is small and constant.

4-1.4 Alternate-Current Motors

In recent years several European CNC manufacturers started to use alternate-current (ac) synchromotors as drives to machine tools. Unlike dc motors, ac motors operate without brushes, thus eliminating one of the main maintenance problems associated with dc motors. The velocity of the ac synchromotor is controlled by manipulation of the voltage *frequency* supplied to the motor, rather than the voltage *magnitude*, as in dc servomotors. The frequency manipulation requires the use of an electrical *inverter*. The inverter contains a dc power supply and a circuit that inverts the resultant dc voltage into ac voltage with a continuously controllable frequency. Inverters have been very costly, and their size was huge compared with dc power amplifiers, which are required to control the voltage magnitude applied to dc motor drives. European manufacturers have recently found a method to build cost-effective small-size inverters, thus enabling the use of ac motors as machine tool drives.

4-2 FEEDBACK DEVICES

In a closed-loop system, information about the output is fed back for comparison with the input. Thus, feedback is required to close the loop. Feedback elements in NC and robot systems are usually rotary devices which are directly coupled to the machine leadscrews, and the robot axes, and provide position and velocity signals.

4-2.1 Encoders

An incremental rotary encoder is one of the most commonly used feedback devices in robots and NC systems. The rotary encoder is a shaft-driven device delivering electrical pulses at its output terminals. The pulse frequency is directly proportional to the shaft speed. The encoder contains a glass disk mounted on the shaft and marked with a precise circular pattern of alternating clear and opaque segments on its periphery, as is shown in Fig. 4-6. A fixed source of light is provided on one side of the disk, and a photocell is placed on its other side. When the disk rotates, light is periodically

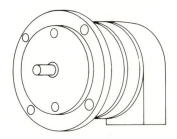

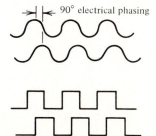

Figure 4-6 Incremental encoder with two-channel output.

permitted to fall on the photocell, which consequently produces a sinusoidal output signal in the millivolt range. This signal is amplified and fed to a Schmitt-trigger circuit, which converts it to a square wave with suitable rise and fall times.

The direction of rotation may be sensed by using an encoder with two photocells reading the same disk. The photocells are arranged so that their outputs have a 90° shift to each other, as shown in Fig. 4-7. The direction of rotation can be determined by external logic circuitry, fed by these two sequences of pulses. An additional index pulse can be available when a separate zone containing only a single clear section is provided

90° electrical phasing

Figure 4-7 (*a*) Phase-shifted output produced by a two-channel encoder; (*b*) square-wave outputs are achieved by applying a Schmitt-trigger shaper.

on the disk. The index pulse can serve as a zero reference position, and is very useful in robot systems.

One disadvantage of using incremental encoders to determine position is the possibility of incorrect data resulting from false counts being generated by noise, transients, or other outside disturbances. Gross errors can also result from power interruption. Those errors are eliminated by using absolute encoders. Absolute rotary encoders use a multiple-track disk which defines the shaft position by means of a binary word (Fig. 4-8) or another code, such as the Gray code. The reading system employs

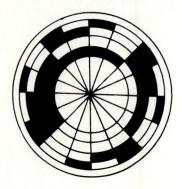

Figure 4-8 Binary code disk of absolute encoder.

a lamp and photocells to detect the light which passes through the transparent portions of the disk. A photocell is provided for each track on the disk. The output from all cells gives the actual shaft position in coded form.

Absolute encoders are most suitable for robots and rotary tables but are only occasionally used in machine tool applications.

4-2.2 Resolvers

Resolvers have the same general construction features as small ac motors. They consist of a rotor and a stator, both having two windings at 90° to one another, as illustrated in Fig. 4-9.

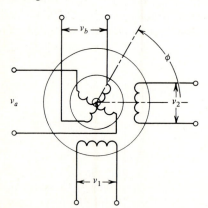

Figure 4-9 Schematic diagram of resolver.

If an ac voltage is applied to one of the stator coils, a maximum voltage will appear at a rotor coil when those two coils are in line, and the voltage will be zero for a $\pm 90°$ shift. As the shaft is turned, the voltage induced in one rotor coil follows a sine wave and the voltage induced in the other follows a cosine wave. Similarly, the ac voltage can be applied to one of the rotor coils, resulting in a sine and cosine of the angular position of the rotor at the two stator winding outputs.

In NC systems the resolver is used as a shaft position measuring device and is directly coupled to the leadscrew of the machine tool. The two windings of the stator are excited by sinusoidal signals equal in frequency and amplitude, but displaced by 90° from each other. That is

$$v_1(t) = V \cos \omega t \qquad (4\text{-}19a)$$

$$v_2(t) = V \sin \omega t \qquad (4\text{-}19b)$$

The rotor outputs consist of two components

$$v_a = v_1 \sin \phi + v_2 \cos \phi \qquad (4\text{-}20a)$$

$$v_b = v_1 \cos \phi - v_2 \sin \phi \qquad (4\text{-}20b)$$

Substituting Eqs. (4-19) into Eqs. (4-20) yields

$$v_a = V \sin (\omega t + \phi) \qquad (4\text{-}21a)$$

$$v_b = V \cos (\omega t + \phi) \qquad (4\text{-}21b)$$

In NC applications only one of the rotor windings is used, and it produces the feedback signal v_a. The phase angle, which is contained in the feedback signal, depends on the angular position of the rotor shaft. Note that if the rotor is rotated 90°, for example, the phase shift of the rotor winding output voltage is 90° from the reference.

In the case where the rotor is continuously rotating with an angular velocity ω_0, the feedback signal is

$$v_a = V \sin [(\omega + \omega_0)t + \phi_0] \qquad (4\text{-}22)$$

where ϕ_0 is the cumulative angular value before reaching the steady state. The sinusoidal voltage of the rotor is converted into a square wave and compared with a square wave corresponding to the stator voltage, which is used as the reference signal. The comparison is performed by a phase-comparator circuit, which detects the phase shift between the reference and the feedback signals and generates a corrective signal.

4-2.3 The Inductosyn

The inductosyn is a precision measuring device developed by the Farrand Controls. Its principle of operation is similar to that of a resolver with a very large number of stator poles P (e.g., $P = 144$), rather than two, and with only one rotor coil. When a single-phase ac voltage is applied to the rotor windings of a resolver, the voltage output from the two-phase stator windings is proportional to the sine and cosine of the angular position of the rotor relative to the stator as shown in Fig. 4-10a. In an inductosyn, due

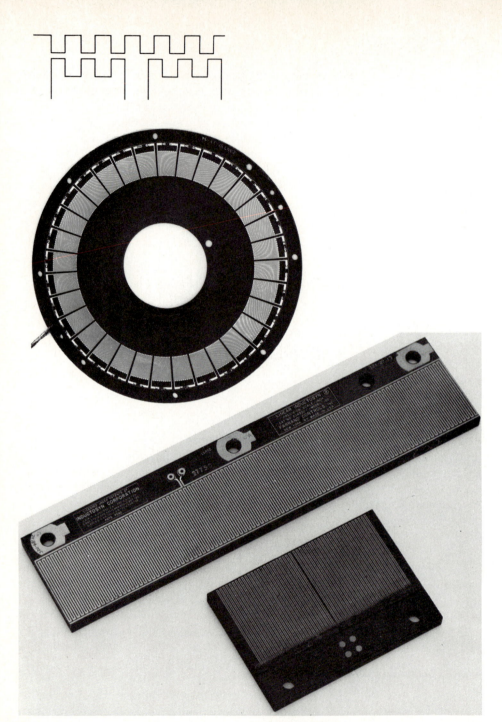

Figure 4-10 (*a*) Two windings, phase shifted by 90° produce two sinusoidal waves in inductosyn; (*b*) stator of rotary inductosyn; (*c*) stator (scale) and rotor (slider) of linear inductosyn. *(Courtesy of Farrand Controls.)*

to the large number of poles, the stator winding outputs are a multiple-angle function and complete one cycle for an angular rotation equal to twice the pole spacing. For a P-pole inductosyn the stator voltages are

$$v_1 = v \, \sin\left(\frac{P\phi}{2}\right) \qquad (4\text{-}23a)$$

$$v_2 = v \, \cos\left(\frac{P\phi}{2}\right) \qquad (4\text{-}23b)$$

where ϕ is the angular position of the rotor relative to the stator and V is the voltage supplied to the rotor. The stator windings are arranged in groups to permit the required 90° phase difference between them as shown in Fig. 4-10b. When the sense of direction is required, the two stator windings are used.

Inductosyns are also made in a linear form for direct measurement of the position of the machine tool table. A linear inductosyn is shown in Fig. 4-10c. The poles are spaced at fixed linear intervals of $\ell/2$ length-units. This provides a linear displacement of ℓ length-units (e.g., $\ell = 0.1$ in, or 2 mm) per cycle. The stator consists of two separate windings, displaced by $\ell/4$ length-units. The relation between the induced voltages at the stator and the displacement is

$$v_1 = v \, \sin \frac{2\pi x}{\ell} \qquad (4\text{-}24a)$$

$$v_2 = v \, \cos \frac{2\pi x}{\ell} \qquad (4\text{-}24b)$$

where x is the linear displacement, and v is the ac voltage supplied to the rotor. Only the amplitudes of v_1 and v_2 change. The phase remains constant, while the amplitude is a function of relative displacement x/ℓ. There is a unique pair of induced voltages v_1 and v_2 for every location within one cycle ℓ. Thus by measuring these voltages, we can subdivide the accurately known cycle interval with high precision.

When the stator windings of an inductosyn transducer are excited by constant amplitude carriers in 90° phase shift, the resulting output signal is a constant amplitude signal that undergoes a continuous phase shift of 360° for each displacement of one cycle length. The constant amplitude signal is easily converted to square wave form for use in NC applications. In the rotary inductosyn the rotor is directly mounted on the leadscrew. In the linear form the stator is fixed to the bed of the machine tool, and the rotor to the table; thus the rotor moves parallel to the stator. The output from the rotor is fed into an input of a phase detector and compared with one of the stator voltages. The resulting error signal from the phase detector is applied to the machine drive and provides the required movement of the machine table.

4-2.4 Tachometers

In order to obtain a precise control of the servomotor speed, the actual speed must be measured and compared with the required one. The actual speed may be measured in terms of voltage by a small PM dc generator, or *tachometer*, coupled to the motor shaft.

The difference between a command voltage proportional to the desired speed and the tachometer voltage may be used to actuate the motor in such a way as to tend to eliminate the error. Negative-feedback closed-loop control is thereby achieved.

The output voltage V_p of the dc tachometer is given by Eq. (4-4):

$$V_p = K_p\omega \tag{4-25}$$

where K_p is the tachometer constant, which is mainly dependent on the magnetic strength of the permanent magnet. In practice the tachometer voltage is fed back through an operational amplifier, which enables the adjustment of K_p.

A simple control loop containing a dc servomotor, amplifier, and tachometer is shown in Fig. 4-11. This loop is referred to as the machine drive unit and is contained

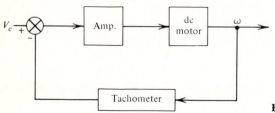

Figure 4-11 Typical machine drive unit.

in the position control of each axis of contouring systems. The difference between the command voltage V_c and the tachometer voltage is the error signal e:

$$e = V_c - V_p \tag{4-26}$$

This error signal is fed to the power amplifier, which produces a voltage V:

$$V = K_a e \tag{4-27}$$

where K_a is the power amplifier gain. The voltage V is applied to the motor input terminals. Combining Eqs. (4-12) and (4-25) through (4-27) yields the closed-loop system Laplace equation:

$$\omega(s) = \frac{\alpha K_a K_m V_c(s) - (\alpha R K_m / K_t) T_s(s)}{1 + s\alpha\tau_m} \tag{4-28}$$

where α is an attenuation factor:

$$\alpha = \frac{1}{1 + K_a K_m K_p} \tag{4-29}$$

Comparison of Eq. (4-12) with Eq. (4-28) shows that the form of the system equations is the same and that the effect of the tachometer is

1. To reduce the time constant (since $\alpha < 1$, then $\alpha\tau_m < \tau_m$)
2. To reduce the effect of the load torque
3. To reduce any nonlinearities of the amplifier gain K_a
4. To facilitate the overall gain adjustment by adjusting the gain of the operational amplifier attached to the tachometer

Example 4-4 The motor in Example 4-1 is driven by a power amplifier with a gain of $K_a = 10$. The loop is closed, as is shown in Fig. 4-11. The tachometer gain is 17.25 V/1000 rpm.

(a) Calculate the time constant of the machine drive unit.
(b) Derive the steady-state equation of the machine drive unit.
(c) Calculate the input voltage V_c which causes the motor to rotate the 985 rpm at the no-load condition.
(d) Calculate the change in speed for a full-load (120 lb·in) condition.

SOLUTION

(a) The motor gain K_m is

$$K_m = \frac{1}{K_v} = \frac{60}{0.824 \times 2\pi} = 11.6 \text{ rpm/V}$$

From Eq. (4-29) the attenuation factor is

$$\alpha = \frac{1}{1 + (10 \times 11.6 \times 17.25/1000)} = \frac{1}{3}$$

The motor time constant is $\tau_m = 13$ ms; the drive system time constant is $\alpha\tau_m$, or 4.33 ms.

(b) At steady state the time constant τ_m does not affect the system performance, and consequently it can be eliminated from Eq. (4-28):

$$\omega = \alpha K_a K_m V_c - \frac{\alpha R K_m T_s}{K_t} \tag{4-30}$$

The variables ω, V_c, and T_s in Eq. (4-30) can be either Laplace or time variables.

(c) At no-load, $T_s = 0$, and the voltage is

$$V_c = \frac{\omega}{\alpha K_a K_m} = 25.5 \text{ V}$$

(d) For a loaded motor the change in speed is

$$\frac{\alpha R K_m T_s}{K_t} = 26 \text{ rpm}$$

which is one-third of the speed change in the open-loop case (see Example 4-1).

4-3 COUNTING DEVICES

The transfer of information in digital form requires special circuits which are called digital circuits or logic circuits. The logic circuits are able to store data and instructions, receive new data, perform arithmetic operations, and transfer the results. They operate at two distinct voltage levels, corresponding to the 0 and 1 values of the boolean algebra variables. These levels are known as H (high), the more positive voltage, and L (low), the zero or less positive voltage.

The basic logic circuits may be divided into two groups:

1. Gates, in which the resulting output depends only on the present input
2. Storage elements, in which the resulting output depends on both the past and present input signals

The logic gates are devices which perform the arithmetic operations of boolean algebra. Every boolean operation has its corresponding gate: AND, OR, NAND, NOR, and EXCLUSIVE-OR. The gates are not discussed here, and it is assumed that the reader is familiar with their principles. Note that gates have a common characteristic: their output is a function of the present state of their inputs.

4-3.1 Flip-Flops

A different kind of logic circuit is an element capable of storing or memorizing information. In memory circuits, the output depends not only on the present level of inputs, but also on the past, or prior, sequence of inputs. The logic state of these circuits is changed by pulses, rather than by logic levels as in gates. A pulse is characterized by a temporary change in the logic level for a short period of time.

The basic memory circuit is the *flip-flop (FF)*, which is a binary storage device that has two distinct stable states, and it remains in one of them until it is directed to change it. The change between the two states is done by means of two inputs, termed *set* and *reset*. Whenever a bit of 1 logic level is stored, the device is said to be set. The operation which stores a 0 bit in a flip-flop is called the reset, or clear, operation, and the flip-flop is said to be in the reset state. For sensing the state, the flip-flop is provided with two outputs Q and Q'. When Q is at 1 logic level, Q' is at 0, and vice versa.

The most commonly used types of binary storage are the RS, JK, and T flip-flops. The T, or trigger, flip-flop (TFF), has only one input, denoted by T or CP (clock pulse). It changes its state each time that the input is triggered by a pulse or by the falling edge of the input signal. *Falling edge* means the transient change between 1 to 0 logic level.

Flip-flops are the basic elements of registers and counters which are used in NC systems. Registers consist of groups of identical flip-flops and are used to store binary information. For example, the binary number 1001 can be represented by a setup of 4 flip-flops, which is termed a 4-bit register. A 4-bit register can store a maximum of 16 different binary words.

4-3.2 Counters

The logic circuit of Fig. 4-12 consists of three TFFs, where the output of each one is connected to the CP input of the next stage. The falling edge at the output of each FF is used as the trigger pulse to the next one. Assume that the FFs have been reset, namely their initial state is $Q_a = Q_b = Q_c = 0$. The first input pulse triggers the FF_a and

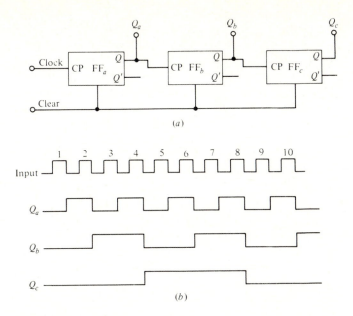

Figure 4-12 A 3-bit binary counter. (*a*) Logic diagram, (*b*) waveforms.

changes its state to $Q_a = 1$. The next pulse changes back the output to $Q_a = 0$, thus providing a falling edge to FF_b and changing its state to $Q_b = 1$. Similarly, FF_c will be triggered only when FF_b returns to the 0 state, and this will occur after four incoming pulses at the input terminal. The generated waveforms are shown in Fig. 4-12, and the FFs' operational states are summarized in Table 4-1.

Table 4-1 Binary counter states (Fig. 4-12)

Q_c	Q_b	Q_a	Pulse no.
0	0	0	Clear
0	0	1	1
0	1	0	2
0	1	1	3
1	0	0	4
1	0	1	5
1	1	0	6
1	1	1	7
0	0	0	8

The reader will have observed that when multiplying the Q_a value by 2^0, that of Q_b by 2^1, and that of Q_c by 2^2, the circuit becomes a binary pulse counter. It is capable of counting, in the binary number system, from 0 to 7. After the count of 7 is reached, the next input pulse will reset the counter to the 0 state. A counter of n flip-flops is capable of being in 2^n distinct states, which means that it can count from 0 through $2^n - 1$. Such a counter is called an n-*bit binary counter*.

In general, counters may be categorized in three ways:

1. According to the base of counting: binary, decimal, etc.
2. According to counting direction: up, down, and up-down
3. Asynchronous and synchronous counters

The base. The binary counter which has been described above is the simplest one. Using n FFs and additional gates, a counter of any base N (also denoted a modulo-N counter) may be constructed, where

$$N \leq 2^n$$

Counters may be used as frequency dividers. A modulo-N counter can be used to divide an incoming frequency by N. The divided frequency is available from the output of the last stage, as is shown, for example, in Fig. 4-12.

Direction of counting. The 3-bit binary counter which was described above is an up-counter, or a forward-counter. An up-counter is a circuit that counts from 0 to some quantity q, determined by the number of flip-flops employed and their intercoupling. It has at least two inputs: one input denoted as count or clock for the incoming pulses, and a second input denoted as reset or clear for resetting the flip-flops. The counter can produce a carry pulse after a count of q has been reached.

A down-counter is a counting circuit which successively reduces the count from some quantity toward 0 in increments of 1. A binary down-counter can be constructed by coupling the Q' output of each FF to the CP input of the next one, instead of coupling the Q output, as is done in an up-counter. The Q output of each FF is still used as the readout output.

A logic circuit which is capable of counting in both forward and reverse directions is called an up-down (U/D) counter. Two main types of U/D counters, which differ by their inputs, exist:

1. A counter with a single clock input for the incoming pulses, and a controlling channel input. Direction of counting is steered from the latter where a 0 logic level will cause a down count and a 1 logic level will accomplish an up count.
2. A counter with two clock inputs: up and down. The direction of counting is determined by which input is pulsed, while the other is at constant logic level.

U/D counters are usually fully programmable; that is, the outputs may be preset to any desired state by entering the data through preset inputs and providing a load instruction. The output changes to agree with the preset inputs independently of the count pulses.

Synchronous and asynchronous counters. Synchronous counters are those counters in which all the FFs receive the same clock at their CP input. This results in all FFs changing their state simultaneously, when so instructed by the steering logic. This mode of operation eliminates counting spikes, which are normally associated with asynchronous counters.

Asynchronous, or ripple-clock, counters differ from synchronous counters in that the FFs do not have a common clock. Each CP input is connected to the output of the previous FF, as is shown in Fig. 4-12.

4-3.3 Decoders

A decoder circuit is often used in conjunction with a counter to provide a method of decoding the counter contents. This is required, for example, when an axis of the NC system should be decelerated before reaching the end of a segment of a workpiece. The down-counter associated with the position control represents the actual distance from the endpoint. The contents of the counter is decremented by the pulses from the encoder until reaching a predetermined value which is automatically identified by the decoder. The decoder, in turn, energizes a deceleration circuit which slows down the corresponding axis of motion.

A 6-decoder connected to a 4-bit counter is shown in Fig. 4-13. The decoder is actually one AND gate. The output of the gate is at 1 logic level only if all its inputs are at 1, and this happens in Fig. 4-13, when the flip-flops are at 0, 1, 1, and 0 states, respectively (i.e., a binary 6). By modifying the wiring between the gate and the counter, a decoder of any number can be constructed.

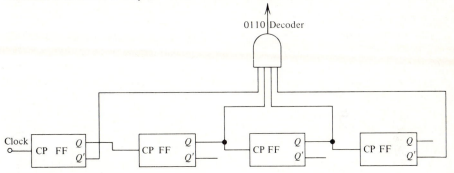

Figure 4-13 A 4-bit counter with a 6-count decoder.

4-4 DIGITAL-TO-ANALOG CONVERTERS

A digital-to-analog converter (DAC) is a device that converts binary information to a proportional analog voltage. In a typical NC system a binary word could be generated in an up-down counter which is used as the loop comparator. The counter is fed by two sequences of pulses: the reference and the feedback pulses from the encoder. The pulse number difference between these two sequences is a binary word, which represents the instantaneous position error, and is converted by the DAC into an analog voltage. The converted output voltage is fed to an amplifier whose output drives a dc motor in a direction which decreases the position error.

Figure 4-14 shows a DAC which takes a binary word, stored in an up-down

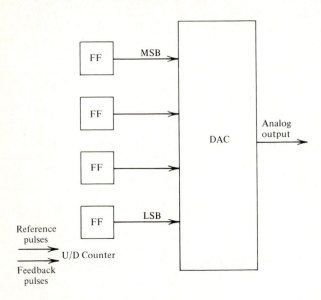

Figure 4-14 Binary word conversion by a DAC.

counter, and converts it into an analog voltage. Each FF drives one input of the DAC. A logic level 1 in a particular input increases the voltage by a magnitude which is proportional to the weight of that input.

The analog voltage versus the digital value of the binary word input is shown in

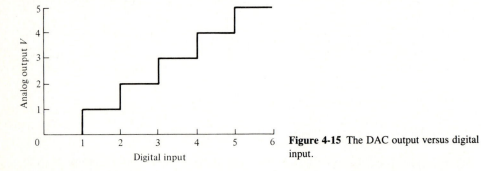

Figure 4-15 The DAC output versus digital input.

Fig. 4-15. Since the input refers only to discrete values, the equivalent analog output is also defined in discrete steps. If the word size is bigger, the steps are smaller and the output looks more continuous.

A basic DAC circuit contains a resistor network, which is the heart of the DAC, and an operational amplifier. Commercial models also contain buffer transistor switches and reference voltage sources. Two types of networks are commonly used: the *weighted resistor* and the *resistor ladder* networks. The accuracy characteristics of the networks are similar and are governed by the precision of the resistors and the reference voltage sources used in the DAC.

4-4.1 Weighted Resistor Network

Figure 4-16 shows a 4-bit DAC, which uses a weighted resistor network. Conversion of an n-bit binary word requires a network of n resistors with n different resistance values. The value of each resistor is inversely proportional to the weight of the particular bit in the binary word.

The operation of the DAC is based on ar. algebraic summation with an operational amplifier. In the process of summation each of the summed variables is multiplied by a constant equal to the respective ratio of feedback to input resistance in accordance with the equation

$$V_o = -\frac{R_f}{R_i} V_i \tag{4-31}$$

where V_i is the input voltage, and V_o is the output voltage. The general relationship for

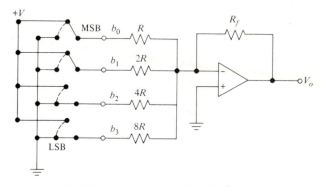

Figure 4-16 DAC with weighted resistor network.

algebraic summation of n input voltages is then

$$V_o = -\left(\frac{R_f}{R_1} V_1 + \frac{R_f}{R_2} V_2 + \cdots + \frac{R_f}{R_n} V_n\right) \tag{4-32}$$

In the weighted network the value of the resistors is that of binary system, with the least-significant bit (LSB) resistance being 2^n times that of the most-significant bit (MSB). Each of the inputs in Fig. 4-16 is energized either by a voltage V, associated with a logical 1, or with a zero voltage, associated with a logical 0. It is assumed that each switch is independently moved from the ground to the $+V$ terminal when a 1 appears at the corresponding digital input. In the commercial DAC the mechanical switches are actually transistor switches with a fast response time.

Substitution of the resistor values in Eq. (4-32)—namely, $R_1 = R$, $R_2 = 2R, \ldots, R_n = nR$—yields the DAC output voltage equation

$$V_o = -V\frac{R_f}{R}\sum_{i=0}^{n-1}\frac{b_i}{2^i} \tag{4-33}$$

where b_i is the logic state of ith bit and can be either 1 or 0. The change in output voltage per bit is called the *gain* of the DAC and is given by

$$K_c = \left|\frac{R_f V}{R\,2^{n-1}}\right| \tag{4-34}$$

Example 4-5 The terminal input voltage of a 4-bit weighted resistor DAC is 2V. The network resistors are 15K, 30K, 60K, and 120K Ω; the feedback resistor is 40K Ω. Calculate The maximum absolute output voltage and the DAC gain.

SOLUTION The maximum output voltage according to Eq. (4-33) is

$$\left|V_{max}\right| = \frac{2\times40}{15}\left(1 + \frac{1}{2} + \frac{1}{4} + \frac{1}{8}\right) = \frac{80}{15}\times\frac{15}{8} = 10\text{ V}$$

According to Eq. (4-34) the gain is

$$K_c = \frac{40\times2}{15\times8} = \frac{2}{3}\text{ volts per bit}$$

4-4.2 Resistor Ladder Network

The weighted resistor DAC requires a wide range of resistor values when the word size n is large. This imposes high precision requirements upon the circuit, which increases the cost significantly. The resistor ladder network, which is shown in Fig. 4-17, overcomes this problem. Only two resistance values are required: R and $2R$, and this can be achieved with very high accuracy. The significant feature of this network is that its output resistance is always R, and it is independent of n. Again, the input signals are represented by switches, where a logical 0 means zero voltage and a logical 1 is represented by $+V$ volts. Practically, the b_i input terminals are coupled to a register or a counter.

The principle of superposition can be used to determine the output voltage. Each bit is considered separately, while the other bits are assumed to be zero. The voltage component produced by each bit is calculated, and then the algebraic sum of all the voltage components is determined. Since the output resistance of this network is R, it can be shown [6] that the output voltage component of the bit b_i is

$$V_i = -\frac{VR_f}{2^iR} \tag{4-35}$$

By the principle of superposition, the total output V_o is the sum of the individual

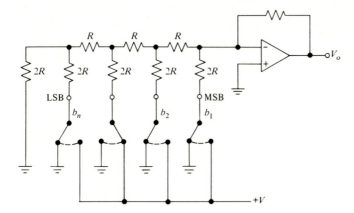

Figure 4-17 DAC with resistor ladder network.

output components

$$V_o = -V \frac{R_f}{R} \sum_{i=1}^{n} \frac{b_i}{2^i} \tag{4-36}$$

where b_i represents the logic state of the ith bit and can be either 0 or 1. If the resistor values and voltage sources are accurate, the output voltage is a linear function of the number to be converted. The gain of the resistor ladder DAC, in terms of volts per bit, is given by

$$K_c = \left| \frac{VR_f}{2^n R} \right| \tag{4-37}$$

Commercially available DACs usually employ the resistor ladder network.

BIBLIOGRAPHY

1. Bakel, J. F.: Hydraulic servo drive for Mark Century controls, General Electric Company, Control Department, Waynesboro, Virginia, 1970.
2. Fitzgerald, A. E., and C. Kingsley: "Electric Machinery," McGraw-Hill Book Company, 2d ed., New York, 1961.
3. Gibson, J. E., and F. B. Tuteur: "Control System Components," McGraw-Hill Book Company, New York, 1958.
4. Graeme, J. G., G. E. Tobey, and L. P. Huelsman: "Operational Amplifiers, Design and Applications," McGraw-Hill Book Company, New York, 1971.
5. "Inductosyn—Principles and Applications," Farrand Controls Inc., 1968.
6. Koren, Y., and J. Ben-Uri: "Numerical Control of Machine Tools," Khanna Publishers, Delhi, 1978.
7. Korn, G. A., and T. M. Korn: "Electronic Analog and Hybrid Computers," McGraw-Hill Book Company, New York, 1964.

8. Kopperschläger, F. D.: "Über die mechanischen Übertragungselemente and numerisch gesteuerten Werkzeugmaschinen," Diss. Rhein.-Westf. Tech. Hochsch., Aachen, 1969.

9. Meyringer, V.: Elektrische Schrittmotoren, Diss. Rhein.-Westf. Techn. Hochsch., Aachen, 1968.

10. Stepping Motors and Incremental Servos, *Contr. Eng.* reprint no. 944, 1970.

11. Suskind, A. K.: "Notes on Analog-Digital Conversion Techniques," Wiley, New York, 1957.

PROBLEMS

4-1 The motor and the valve constants in a hydraulic system are $K = 2000$ rad/gal, and $K_v = 0.5$ gal/min·V, respectively. If a signal of 5 V energized the valve, what is the motor speed in revolutions per minute?

4-2 The motor of Example 4-1 was mounted as the axial drive of a CNC system and drives the leadscrew through a gear ratio of 2:1, which reduces the motor speed. If the leadscrew load is 120 lb·in, and its moment of inertia is 0.46 lb·in·s^2, calculate the drive time constant and the steady-state speed. Draw the speed-time diagrams of an unloaded (Example 4-1) and a loaded motor.

4-3 A separately excited dc motor has the following constants:

Armature resistance, $R = 0.5\ \Omega$

Armature inductance negligible

Torque constant, 2 N·m/A

Total moment of inertia of load and armature, $J = 8$ kg·m^2

The load is pure inertia

The motor armature is connected to a source of 110 V at $t = 0$

(*a*) Derive an expression, with numerical values, for the speed in radians per second as a function of time. Sketch the curve.

(*b*) Calculate the steady-state speed and the time required to reach 90 percent of this value.

4-4 The motor of Prob. 4-3 is loaded by a constant torque of 80 N·m after reaching the steady-state speed. Calculate the final change in speed.

4-5 Design a dc motor drive for a milling machine NC table.

Maximum axial velocity, 2500 mm/min

Resolution, BLU = 0.02 mm

Leadscrew pitch, 10 mm (assume leadscrew efficiency 100 percent)

Maximum speed of the motor, 1000 rpm

The feedback device is an incremental encoder which is directly mounted on the leadscrew. It is assumed that the maximum axial force (friction and cutting force) during milling is 400 N and is generated at velocity of 600 mm/min. Calculate:

(*a*) The encoder gain (pulses per revolution) and its maximum frequency (pulses per second).

(*b*) The required gear ratio between the motor and the leadscrew.

(*c*) The maximum static loading torque on the motor (in newton-meters).

(*d*) Derive the general relationship between the axial force F and the motor's torque T. Use the leadscrew pitch (P) and the gear ratio (N) as parameters.

4-6 The pull-in curve (L in Fig. 4-5) of a 200 steps-per-revolution stepping motor can be approximated by the function

$$T = -0.1V^2 + 810$$

where T is the torque in gram-centimeters and V is the velocity in steps per second.

The motor is coupled to a leadscrew of 5-mm pitch, which is connected to a load described by the equation

$$T = 200 + 25v$$

where v is the load velocity in millimeters per second, and T is given in gram-centimeters.

(*a*) Draw a torque-frequency graph of the motor and load.

(*b*) What is the maximum starting frequency to the motor?

4-7 A stepping motor with a step increment of $1.8°$ is directly mounted on a leadscrew of 10-mm pitch, which loads the motor with a constant torque of 160 oz·in. The torque speed characteristic (curve L) of the motor can be approximated by

$$T = 430 - 0.045V$$

where V is given in steps per second.

(a) Calculate the motor speed (in steps per second) required to obtain a velocity of 250 mm/s of linear motion.

(b) Check if the stepping motor is able to drive the NC worktable.

(c) If the moment of inertia is 4.5 oz·in·s^2, what is the maximum allowable acceleration rate of this stepping motor?

4-8 The motor of Prob. 4-3 is loaded with 80 N·m and connected in a closed-loop as shown in Fig. 4-11. The amplifier gain is $K_a = 10$, and the tachometer gain is 0.2 V/(rad/s). The input voltage to the loop is 22 V.

(a) Calculate the time constant of the closed-loop drive system.

(b) Calculate the steady-state speed of unloaded and loaded motor.

4-9 Modify the connections between the decoder and the counter in Fig. 4-13 to obtain a 9-decoder.

4-10 A 10-bit ladder resistor DAC has the following values:

$$V = 5.12V; R = 50\text{K}\Omega$$

(a) Calculate the value of R_f which is required to obtain a gain of 0.01 volts per bit.

(b) What is the maximum possible voltage at the output?

4-11 The input to the $2R$ resistor of a weighted resistor DAC (Fig. 4-16) was disconnected. Draw the DAC output voltage as a function of its input number.

FIVE

INTERPOLATORS FOR MANUFACTURING SYSTEMS

A common requirement of all manufacturing systems is to generate *coordinated movement* of the separately driven axes of motion in order to achieve the desired path of the tool relative to the workpiece. This involves the generation of signals prescribing the shape of the produced part and their transmission as reference inputs to the corresponding control loops. Generation of these reference signals is accomplished by *interpolators*. NC systems contain hardware interpolators which consist of digital circuits, whereas in CNC systems the interpolator is implemented in software.

5-1 DDA INTEGRATOR

Digital differential analyzers (DDA) are a special type of computer in which all variables are represented by digital words, but the solution technique is similar to that used in analog computers. The DDA computer has the advantages of both digital and analog computers, and combines the accuracy of digital computation with the continuity of operation of the analog computer. Consequently it is suitable for digital solution of differential equations in real time.

The fundamental element of the DDA computer is the DDA integrator, which plays a role similar to the operational amplifier in analog computers and forms the basic block for integration. The DDA integrators are interconnected similarly to analog computer setups. The difficulty of transferring n-bit words between integrators is avoided by using an incremental transfer computation method. This means that only increments of variables and their sign bit are transferred, so that only two lines, instead of $(n + 1)$ transmitting lines, are required.

5-1.1 Principle of Operation

Digital integration is basically performed by successive additions using rectangular or trapezoidal approximation methods. The DDAs in NC systems employ the rectangular approximation. Let us assume that a variable is a function of the time t, as illustrated in Fig. 5-1. Digital integration is performed by approximating the area below the curve

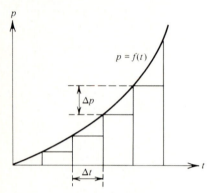

Figure 5-1 Digital approximation of a continuous function.

as a sum of small rectangular areas, each of equal base $\triangle t$. This yields

$$z(t) = \int_0^t p \, dt \cong \sum_{i=1}^{k} p_i \, \Delta t \tag{5-1}$$

The value of z at $t = k\Delta t$ is denoted by z_k, which may be written as follows:

$$z_k = \sum_{i=1}^{k-1} p_i \, \Delta t + p_k \, \Delta t \tag{5-2}$$

or
$$z_k = z_{k-1} + \Delta z_k \tag{5-3}$$

where
$$\Delta z_k = p_k \, \Delta t \tag{5-4}$$

Digital integration is accomplished in three stages. First the value of the ordinate p_k is computed by adding the increment Δp_k to, or subtracting it from, the preceding ordinate:

$$p_k = p_{k-1} \pm \Delta p_k \tag{5-5}$$

Then the integration increment Δz is calculated according to Eq. (5-4), and finally it is added to the previous z according to Eq. (5-3).

The DDA integrator operates in an iterative mode at a frequency f provided by an external clock, where

$$f = \frac{1}{\Delta t} \tag{5-6}$$

During each iteration the operations given by Eqs. (5-4) and (5-5) should be executed. As was noted above, the input and output data between DDA integrators is transferred as 1-bit increments, and therefore the values of Δp and Δz must be either 1 or 0. This is achieved in DDAs by storing the p variable in an n-bit register, or up-down counter, and thus limiting its allowable value to 2^n, namely

$$\frac{p_k}{2^n} < 1$$

The input increment Δp, which is either 1 or 0, is added to the least-significant bit (LSB) of this n-bit register, which is denoted as the p register. The output increment is calculated with the aid of an additional n-bit register, denoted as the q register. At each iteration the variable p is added to the previous contents of q

$$q_k = q_{k-1} + p_k \tag{5-7}$$

If the new value of q is greater than $(2^n - 1)$, which is the maximum possible value with an n-bit word, an overflow Δz is generated. For example, if $n = 3$, $p_k = 6$, and $q_{k-1} = 4$, the binary addition is executed as follows:

$$
\begin{array}{cc}
\begin{array}{rr}
 & q_{k-1} \\
+ & \\
 & p_k \\
\hline
\boxed{\Delta z} & q_k
\end{array}
&
\begin{array}{rr}
 & 100 \\
+ & \\
 & 110 \\
\hline
\boxed{1} & 010
\end{array}
\end{array}
$$

Namely, $q_k = 2$ and $\Delta z = 1$.

A schematic diagram of a DDA integrator is shown in Fig. 5-2. It consists of two

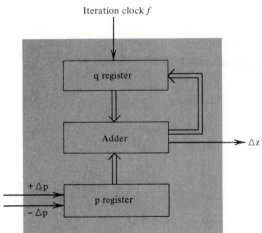

Iteration clock f

q register

Adder → Δz

$+\Delta p$

$-\Delta p$

p register

Figure 5-2 A schematic diagram of a DDA integrator.

n-bit registers, p and q, and one binary adder. During each integration step the new value of p is obtained according to Eq. (5-5), in which Δp is either 1 or 0. The integration itself is performed according to Eq. (5-7) and is executed with the aid of the binary adder which adds the contents of the p and the q registers at each iteration. The adder overflows are the output increments Δz. Mathematically these increments are given by

$$\Delta z_k = 2^{-n} p_k \qquad (5\text{-}8)$$

By combining Eqs. (5-6) and (5-8) they can be written as

$$\Delta z_k = C p_k \, \Delta t \qquad (5\text{-}9)$$

where C is the DDA integration constant, which is defined by

$$C = \frac{f}{2^n} \qquad (5\text{-}10)$$

Equation (5-9) has the same structure as Eq. (5-4).

A symbolic representation of a DDA integrator is shown in Fig. 5-3. The output of

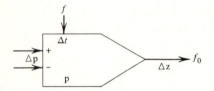

Figure 5-3 A symbolic representation of a DDA integrator.

the DDA is the overflow pulse Δz, which can be connected to the Δp inputs of other DDAs in order to design a desired setup. The DDA itself does not accumulate the Δz pulses; if the integration result is required, an additional counter, which executes Eq. (5-3), must be added.

The average output frequency of the overflow pulses f_0 is obtained from Eq. (5-9):

$$f_0 = \left(\frac{\Delta z}{\Delta t} \right)_k = C p_k \qquad (5\text{-}11)$$

and is proportional to the iteration frequency f and the current value p, and inversely proportional to 2^n where n is the length of the DDA registers. The number n establishes the resolution of the integration process; a larger n results in more accurate integration.

Example 5-1 A simple integration in which p is a constant (i.e., $\Delta p = 0$) is performed with a DDA integrator. Calculate the output Δz at the first 10 integration steps; the DDA contains 3-bit registers which are initially set to $p = 5$ and $q = 0$.

SOLUTION After the first iteration the contents of the q register is 5, according to Eq. (5-7). At the second iteration its contents is calculated with the binary system as follows:

$$
\begin{array}{ccc}
 & q_{k-1} & \qquad\qquad 101 \\
+ & & + \\
 & p & \qquad\qquad 101 \\
\hline
\boxed{\Delta z} \quad & q_k & \qquad \boxed{1} \quad 010
\end{array}
$$

Namely $q_k = 2$ and $\Delta z = 1$. These values can be calculated also with the decimal system according to Eq. (5-7). However, if $q_k \geq 8$ (2^n in the general case), then

$$q_k = q_{k-1} + p_k - 8 \qquad \text{and} \qquad \Delta z = 1 \qquad (5\text{-}12)$$

The results of the first 10 iterations were computed accordingly and are given in Table 5-1. The Δz increments are summed in the $\Sigma \Delta z$ column, which provides the instantaneous integration result. This value and the correct one are shown in Fig. 5-4. It can be seen that the integration accuracy can be improved by setting

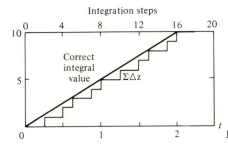

Figure 5-4 Digital integration of a constant.

the q register to a nonzero initial value. Also note that the q values repeat every 8 iterations, or every 2^n steps in the general case.

Table 5-1. Results of Example 5-1.

Step no.	q	Δz	$\Sigma \Delta z$
1	5		0
2	2	1	1
3	7		1
4	4	1	2
5	1	1	3
6	6		3
7	3	1	4
8	0	1	5
9	5		5
10	2	1	6

In NC sytems the machine axes are driven by the Δz pulses; each pulse generates a motion of 1 BLU. The displacement of the axis is actually the integration result and is given by Eq. (5-3). The velocity of the axis in BLUs per second is given by Eq. (5-11).

5-1.2 Exponential Deceleration

A DDA which produces an exponentially decaying frequency is frequently used to decelerate the machine axes of motion before the final stop.

Assume that $p(t)$ is an exponential function:

$$p(t) = p_0 e^{-\alpha t} \tag{5-13}$$

The DDA output frequency is given by Eq. (5-11)

$$\frac{\Delta z}{\Delta t} = Cp_k = Cp_0 e^{-\alpha t} \tag{5-14}$$

By differentiating Eq. (5-13) one obtains

$$dp = -p_0 \alpha e^{-\alpha t}\, dt \tag{5-15}$$

and the corresponding difference equation is

$$-\Delta p = \alpha p_k\, \Delta t \tag{5-16}$$

Comparing Eqs. (5-9) and (5-16) shows that the required exponential function is obtained if

$$-\Delta p = \Delta z \tag{5-17}$$

and the constant C is adjusted such that $C = \alpha$. Equation (5-17) is satisfied if the output Δz of the integrator is connected to its $-\Delta p$ input as is shown in Fig. 5-5.

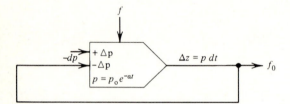

Figure 5-5 Setup of an exponential function.

Table 5-2 Computation of exponential function

Step no.	p	q	Δz	Step no.	p	q	Δz
Initial	15	0					
1	15	15		21	4	8	
2	15	14	1	22	4	12	
3	14	12	1	23	4	0	1
4	13	9	1	24	3	3	
5	12	5	1	25	3	6	
6	11	0	1	26	3	9	
7	10	10		27	3	12	
8	10	4	1	28	3	15	
9	9	13		29	3	2	1
10	9	6	1	30	2	4	
11	8	14					
12	8	6	1	35	2	14	
13	7	13		36	2	0	1
14	7	4	1	37	1	1	
15	6	10		38	1	2	
16	6	0	1				
17	5	5		50	1	14	
18	5	10		51	1	15	
19	5	15		52	1	0	1
20	5	4	1	53	0	0	

Example 5-2 Generate the exponential function $p(t) = 15e^{-t}$ with a DDA that has 4-bit registers.

SOLUTION The setup consists of one DDA as shown in Fig. 5-5. The p register is loaded to 15 and then the clock starts. The contents of the p and q registers are given in Table 5-2. After 53 integration steps $p_k = 0$ and the computation stops. The DDA clock frequency is given by the equation

$$f = \alpha 2^n = 16 \text{ pps}$$

The accumulated z, the decaying value of p, and the DDA output frequency f_0, are plotted in Fig. 5-6.

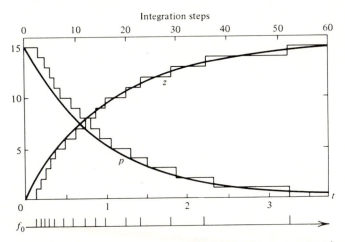

Figure 5-6 The contents of the p register and integration result of $15e^{-t}$.

5-2 DDA HARDWARE INTERPOLATOR

In contouring systems the machining path is usually constructed from a combination of linear and circular segments. It is only necessary to specify the coordinates of the initial and final points of each segment, and the feedrate. The operation of producing the required shape based on this information is termed *interpolation,* and the corresponding electronic unit is the *interpolator.* The interpolator coordinates the motion along the machine axes, which are separately driven, to generate the required machining path. The two most common types of interpolators found in practice are those providing linear and circular facilities. Parabolic interpolators are also available in a few NC systems which are used in the aircraft industry. The DDA interpolator consists of DDA integrators and digital gates which are wired to provide the required interpolation.

5-2.1 Linear Interpolator

The ability to control the movement along a straight line between given initial and final coordinates is termed *linear interpolation.* Linear interpolation can be performed in a plane (2-D), using one or two axes of motion, or in space (3-D), where the combined motion of three axes is required. In this chapter only 2-D linear interpolators are discussed.

The 2-D linear interpolator supplies velocity commands, in pulses per second, simultaneously to two machine axes, and maintains the ratio between the pulse frequencies equal to the ratio between the required incremental distances. For example, consider the case in Fig. 5-7a, where a material has to be cut in a straight path between

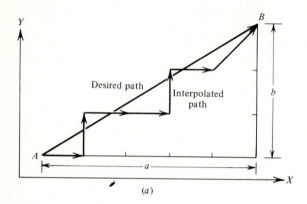

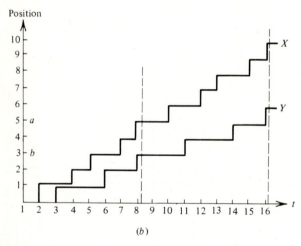

Figure 5-7 Straight-line interpolation.

points A and B. The incremental distances along the X and Y axes are 5 and 3 BLUs, respectively, where each BLU is equivalent to one output pulse. The interpolator must supply 5 pulses to the control loop of the X axis simultaneously with 3 pulses to the Y axis, namely, the ratio between the two frequencies is 5:3. The actual path formed by the interpolator consists of BLU increments, and the largest possible error is smaller than a single BLU. In the actual path this error might be even smaller (neglecting the machine tool inaccuracy) due to the smoothing effect of the servomotors.

A 2-D linear interpolator which consists of DDA integrators is shown in Fig. 5-8.

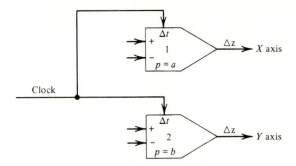

Figure 5-8 Linear DDA interpolator.

Each axis of motion requires one DDA integrator; integrator 1 supplies pulses to the X axis and integrator 2 to the Y axis. The two DDAs are controlled by a common clock, and therefore their operations are performed simultaneously. The required incremental distance of each axis is fed to the p register in the corresponding DDA. The overflow pulses from the Δz outputs are supplied as the command signal to the control loops. These pulses can actuate stepping motors in open-loop systems, where each pulse causes a single step motion, or can be fed as reference to closed-loop systems.

Assume that a straight path AB has to be machined, with increments a and b along the X and Y axes, respectively. The p register of integrator 1 is loaded with a, and that of integrator 2 with b. According to Eq. (5-9) the corresponding outputs are

$$\Delta z_1 = Ca\ \Delta t \qquad (5\text{-}18a)$$

$$\Delta z_2 = Cb\ \Delta t \qquad (5\text{-}18b)$$

where C is given in Eq. (5-10). The ratio of the output frequencies is

$$\frac{\Delta z_1/\Delta t}{\Delta z_2/\Delta t} = \frac{a}{b} \qquad (5\text{-}19)$$

and guarantees that the required path is generated. The instantaneous position of X and Y can be determined by integrating Eqs. (5-18):

$$X = Cat \qquad Y = Cbt \qquad (5\text{-}20)$$

Example 5-3 A linear DDA interpolator with 3-bit registers generates the path given in Fig. 5-7. Determine the output pulses at the first 10 integration steps.

SOLUTION The X axis p register is loaded with 5 and that of Y with 3. Equation (5-7) is performed simultaneously in the two DDAs. The output pulses are generated each time that $q \geq 8$, and then Eq. (5-12) is executed. The results are given in Table 5-3, and are plotted in Fig. 5-7b for the first 16 integration steps. Note that after each 8 integration steps (2^n, in the general case) the q registers are reset, and a new cycle starts. Therefore, the staircase shape illustrated in Fig. 5-7a can continue if required.

Table 5-3 Results of the linear interpolation in Example 5-3

Integration step	Integrator 1		Integrator 2	
	q	Δz_1	q	Δz_2
1	5		3	
2	2	1	6	
3	7		1	1
4	4	1	4	
5	1	1	7	
6	6		2	1
7	3	1	5	
8	0	1	0	1
9	5		3	
10	0	1	6	

So far the interpolator has generated the proper velocity ratio between the two axes. However, the desired velocity along the path should be also maintained. This is achieved by applying an additional DDA integrator (DDA 3 in Fig. 5-12) to control the clock frequency of the first two DDAs.

This additional DDA, which is denoted as the feedrate DDA, has two m-bit registers. Its p register is loaded with the FRN code, which is calculated with the inverse-time method and is given by Eq. (3-4):

$$FRN = 10\frac{V}{L} \tag{5-21}$$

where V is the required feedrate, or velocity, along the path, given in length-units per *minute*, and L is the length of the path given in the same length-units and can be expressed by

$$L = \sqrt{a^2 + b^2} \tag{5-22}$$

The output frequency of this DDA is determined by substituting Eq. (5-21) into Eq. (5-11), which yields

$$f_0 = C \times FRN = \frac{10Vf}{2^m L} \tag{5-23}$$

where f is the clock frequency to DDA 3. The output of DDA 3 is connected to the clock input of DDAs 1 and 2. According to Eq. (5-10) the integration constant of these DDAs becomes

$$C = \frac{f_0}{2^n} = \frac{10Vf}{2^n 2^m L} \tag{5-24}$$

The axial velocities are given by Eqs. (5-18):

$$V_x = \frac{\Delta z_1}{\Delta t} = Ca \tag{5-25a}$$

$$V_y = \frac{\Delta z_2}{\Delta t} = Cb \tag{5-25b}$$

The velocities V_x and V_y in Eqs. (5-25) are given in length-units per second and consequently the actual path velocity V_l in length-units per *second* is

$$V_l = \sqrt{V_x^2 + V_y^2} = C\sqrt{a^2 + b^2} = CL \tag{5-26}$$

Substituting Eq. (5-24) into Eq. (5-26) yields

$$V_l = \frac{10f}{2^{n+m}} V \tag{5-27}$$

The desired path velocity V is obtained (i.e., $V = 60V_l$) if the clock frequency f of DDA 3 is selected as

$$f = \frac{2^{m+n}}{600} \tag{5-28}$$

This is a fixed clock, and its frequency is adjusted by the NC manufacturer. We see that by applying the inverse-time method to calculate the FRN, the on-line computation of L according to Eq. (5-22) is eliminated.

Example 5-4 Calculate the clock frequency of DDA 3 in a NC system with BLU = 0.01 mm. The maximum incremental motion in this system is limited to 600 mm, and the maximum FRN code is 1999.

SOLUTION The maximum incremental motion is 60,000 BLUs, and it can be stored in a 16-bit register ($2^{16} = 65,536$). Therefore, the length of registers p and q in DDAs 1 and 2 is $n = 16$. The maximum FRN number can be stored in an 11-bit register, namely $m = 11$. According to Eq. (5-28) the clock frequency is $f = 223,696$ pps.

5-2.2 Circular Interpolator

The circular interpolator eliminates the need to define many points along a circular arc. Only the initial and final points and the radius are required to generate the arc. In most practical cases the circular interpolation is limited to one quadrant, and therefore both the startpoints and endpoints must be located in the same quadrant of the circle. Larger arcs are divided into successive circular arcs.

For producing a circular arc the following path equation must be satisfied:

$$(X - R)^2 + Y^2 = R^2 \tag{5-29}$$

where R is the radius of the required circle and

$$X = R(1 - \cos \omega t) \tag{5-30a}$$

$$Y = R \sin \omega t \tag{5-30b}$$

The corresponding axial velocity commands are obtained by differentiation:

$$V_x = \frac{dX}{dt} = \omega R \sin \omega t \tag{5-31a}$$

$$V_y = \frac{dY}{dt} = \omega R \cos \omega t \tag{5-31b}$$

Or alternatively written:

$$dX = \omega R \sin \omega t \, dt = -d(R \cos \omega t) \tag{5-32a}$$

$$dY = \omega R \cos \omega t \, dt = +d(R \sin \omega t) \tag{5-32b}$$

The term ωR is the desired feedrate along the circular arc.

The circular interpolator consists of two cross-coupled DDA integrators, as shown in Fig. 5-9. The p registers are loaded with the axial projections of the starting points

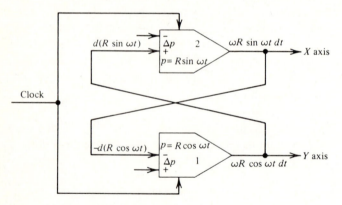

Figure 5-9 Circular DDA interpolator.

i and j (see Fig. 3-5), which are provided by the part program and are defined as

$$i = R \cos \omega t_0 \tag{5-33a}$$

$$j = R \sin \omega t_0 \tag{5-33b}$$

where t_0 is the initial time. The output of the DDAs is given according to Eq. (5-9):

$$\Delta z_1 = CR \cos \omega t \, dt \tag{5-34a}$$

$$\Delta z_2 = CR \sin \omega t \, dt \tag{5-34b}$$

and the value of the p registers is updated by

$$-\Delta p_1 = -d \, (R \cos \omega t) \tag{5-35a}$$

$$\Delta p_2 = d \, (R \sin \omega t) \tag{5-35b}$$

Equations (5-34) have a similar structure to Eqs. (5-32); the integration constant must be adjusted to $C = \omega$. The interpolator operates as follows: DDA 2 is initially loaded to $p = j$ and its output pulses are supplied to the X axis, while DDA 1 is initially loaded to $p = i$ and in turn is connected to the Y axis. The contents $R \sin \omega t$ of the p register of DDA 2 is updated throughout the circular interpolation by the increment $dY = d(R \sin \omega t)$, obtained from the output of DDA 1, as can be seen from Eq. (5-32b). Similarly, the output of DDA 2, which is $-d(R \cos \omega t)$, is connected to the $-\Delta p$ input of DDA 1 for updating the $R \cos \omega t$ value in its p register.

Example 5-5 Assume that a circular quadrant is to be produced with the initial conditions $i = R = 15$ and $j = 0$. The DDA register length is $n = 4$. Calculate the register contents at each integration step.

SOLUTION The step-by-step calculations are given in Table 5-4. It can be seen that since the value of the p register of DDA 1 is initially high, it emits pulses at high frequency. The value of p is gradually reduced by pulses entering the Δp input, and the output frequency drops until p is almost zero. That means that the output frequency of DDA 1 corresponds to a cos ωt function. In contrast, a zero value is initially loaded into the p register of DDA 2, but its p register is gradually filled up with pulses from DDA 1 and the output frequency increases accordingly.

The results in Table 5-4 are plotted in Fig. 5-10. Obviously the smoothness

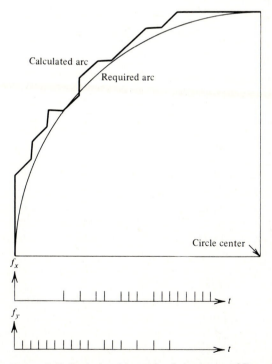

Figure 5-10 Desired and interpolated circular arc of Example 5-5.

of the arc can be improved by increasing the register length n. In practical NC systems n ranges between 14 and 20 and depends on the system resolution and the maximum allowable dimensional word.

Table 5-4 Step-by-step circle calculation

Integration step	DDA 1			DDA 2		
	p	q	Δz	p	q	Δz
0	15	0		0	0	
1	15	15		0	0	
2	15	14	1	1	1	
3	15	13	1	2	3	
4	15	12	1	3	6	
5	15	11	1	4	10	
6	15	10	1	5	15	
7	15	9	1	6	5	1
8	14	7	1	7	12	
9	14	5	1	8	4	1
10	13	2	1	9	13	
11	13	15		9	6	1
12	12	11	1	10	0	1
13	11	6	1	11	11	
14	11	1	1	12	7	1
15	10	11		12	3	1
16	9	4	1	13	0	1
17	8	12		13	13	
18	8	4	1	14	11	1
19	7	11		14	9	1
20	6	1	1	15	8	1
21	5	6		15	7	1
22	4	10		15	6	1
23	3	13		15	5	1
24	2	15		15	4	1
25	1	0	†	15	3	1

† Not sent.

So far we have considered interpolation of a clockwise circular arc in which the incremental direction of motion of both X and Y is positive. However, there are eight different arcs which can start at a point, as illustrated in Fig. 5-11. The one which has been discussed is arc a, but actually arcs b, c, and d also have a similar structure and are generated by the same interpolator. The difference is in the control unit, where the instructed direction of motion varies according to the sign of the incremental dimension words of the part program. For example, if the direction of X is reversed, arc b rather than a is generated with the same interpolator. On the other hand arcs e, f, g, and h belong to another group and are generated by a different interpolator. Again, the specific arc is determined in the control unit and depends on the sign of the dimension words. Note that while an arc of one group is generated in a clockwise direction (e.g., arc a), its counterpart which lies in the same quadrant (arc e) is generated in a counterclockwise direction.

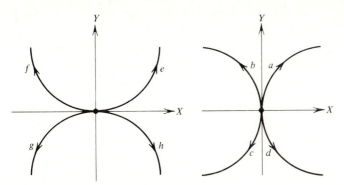

Figure 5-11 Eight possible circular arcs.

The interpolator which generates the arcs of the second group has a structure similar to the one in Fig. 5-9, but the sign at the DDAs input is reversed. Namely, the output of DDA 1 is fed to the $-\Delta p$ input of 2, and the output of DDA 2 is connected to the $+\Delta p$ of DDA 1. The p registers of the DDAs are initially loaded according to Eqs. (5-33), which provide different initial conditions for the two types of circular interpolators. For example, to generate an arc of type a, b, c, or d from the origin in Fig. 5-11, the p registers of the DDAs are loaded as follows:

$$p_1 = i = R \qquad p_2 = j = 0$$

where R is the radius of the required arc. To generate an arc of the group e, f, g, or h, the initial conditions are

$$p_1 = i = 0 \qquad p_2 = j = R$$

The values of i and j are always provided by the part program.

5-2.3 Complete Interpolator

A complete interpolator which is capable of generating both a linear and one type of circular interpolation is shown in Fig. 5-12. The part programmer specifies the type of interpolation through the g word. A network of digital gates at the integrator outputs controls the direction of the DDA overflow pulses. This network also connects the Δz outputs of the integrators to the required Δp inputs in circular interpolation.

The velocity (i.e., feedrate in machine tools) along the path is controlled by DDA 3 in Fig. 5-12. Its p register is loaded with the FRN code, which is calculated by the inverse-time method which was introduced in Chap. 3. For a linear interpolator it was shown that if the clock frequency of this DDA is adjusted according to Eq. (5-28), the required velocity along a straight path is achieved. The same clock also ensures that the required velocity around a circular arc is obtained.

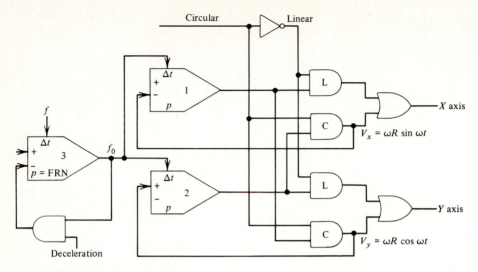

Figure 5-12 Complete DDA interpolator.

In circular arcs, according to Eq. (3-5), the FRN code is given by

$$\text{FRN} = 10\frac{V}{R} \tag{5-36}$$

where V is the required velocity around the arc given in BLUs per minute and R is the radius of the arc in BLUs, given by

$$R = \sqrt{i^2 + j^2} \tag{5-37}$$

where i and j are defined in Eqs. (5-33). According to Eq. (5-11) the output frequency f_0 of DDA 3 is

$$f_0 = \frac{10Vf}{2^m R} \tag{5-38}$$

where f is the interpolator clock frequency given by Eq. (5-28) and m is the number of bits in the p and q registers of DDA 3. The output frequency f_0 is fed to the clock input of DDAs 1 and 2. The output velocities of these DDAs are given in terms of BLUs per second by Eqs. (5-34):

$$V_y = \frac{\Delta z_1}{\Delta t} = CR \cos \omega t \tag{5-39a}$$

$$V_x = \frac{\Delta z_2}{\Delta t} = CR \sin \omega t \tag{5-39b}$$

where C is defined in Eq. (5-10):

$$C = \frac{f_0}{2^n} \tag{5-40}$$

and n is the number of bits in the p and q registers of DDAs 1 and 2. Substituting Eqs. (5-28) and (5-38) into Eq. (5-40) yields the integration constant

$$C = \frac{V}{60R} = \frac{V_\ell}{R} \tag{5-41}$$

namely, the integration constant becomes equal to the angular velocity ω. A further substitution of Eq. (5-41) into Eq. (5-39) gives

$$V_x = V_\ell \sin \omega t \qquad V_y = V_\ell \cos \omega t \tag{5-42}$$

which are the axial velocities required to generate the desired circular arc.

Deceleration may be accomplished by providing a negative feedback to DDA 3, as is shown in Fig. 5-12. The DDA output is connected to its $- \Delta p$ input, and an exponentially decaying output frequency is thus produced. This frequency reduces the rate of iterations in DDAs 1 and 2, and consequently the velocity of the two axes is simultaneously decreased.

5-3 CNC SOFTWARE INTERPOLATORS

CNC is attracting increasing attention in manufacturing. With CNC, a minicomputer or a microcomputer is provided as part of the controller to perform the basic NC functions. These naturally include data processing, feedrate calculations, and interpolation between data points, leaving only the position- and velocity-control loops to the hardware controller. Interpolation is performed by means of a special routine which generates command signals for each segment of the produced part based upon the initial and final points and the type of curvature of the segment. Typical interpolators are capable of generating linear, circular, and occasionally parabolic paths. Elliptic interpolation is inapplicable in NC of machine tools but may be useful in other manufacturing systems such as laser-beam cutters.

Basically there are two types of CNCs: the reference-pulse and the sampled-data systems. In the reference-pulse system, the computer produces a sequence of reference pulses for each axis of motion, each pulse generating a motion of one BLU. The accumulated number of pulses represents position, and the pulse frequency is proportional to the axis velocity. These pulses can either actuate a stepping motor in an open-loop system, or be fed as a reference to a closed-loop system (see Fig. 1-8). With the sampled-data technique, the control loop of each axis is closed through the computer itself, which generates reference binary words. These two CNC types require distinct interpolation routines in the control program to generate their corresponding reference signals (pulses or binary words).

The overall design of a CNC system first requires the selection of the appropriate control technique (reference-pulse or sampled-data) and the optimal setting of the control-loop parameters. Subsequently, the appropriate interpolator routines must be written. The reference-pulse interpolators are simpler to program, but there is a restriction on the maximum axis velocity imposed by the interpolation execution time. Therefore, this method is not suitable for manufacturing systems requiring high axial velocities. By contrast, the maximum velocity in sampled-data systems is not limited by the computer, but the interpolator routine is more complex and the control-loop gain is lower, which results in larger axial position errors.

Reference-pulse interpolators are preferably written in assembly language in order to improve the computing speed since the maximum attainable axis velocity is proportional to the computing speed of the interpolator. A floating-point technique should be avoided, where possible, in favor of a routine using basic assembler instructions such as add, compare, shift, etc.

All reference-pulse interpolators are based upon an iterative technique controlled by an interrupt clock. At each interrupt, a single iteration of the routine is executed, which in turn can provide an output pulse. The maximum axis velocity is proportional to the maximum attainable interrupt frequency, which, in turn, depends on the execution time of the interpolator algorithm.

Three reference-pulse methods are discussed in the literature [12]:

1. Software DDA method, which is based on a simulation of hardware DDA
2. Stairs approximation method, which is based on proceeding along the segment by BLU steps either in the X or in the Y direction
3. Direct search method, which is based on a minimum error criterion

In many manufacturing systems the overriding factor in selecting the interpolation method may be the uniformity of velocity along the path. In a laser-beam cutter, for example, the velocity prescribes the kerf (width of cut), and, consequently, velocity variations along the path also affect the accuracy of the produced part. In milling and turning velocity variations cause feed variations which affect the surface finish of the part. It has been proved [12] that a uniform velocity around a *circular* arc is obtained only with the software DDA method. With the other methods large variations in the velocity (up to 41 percent) can occur along the circular path. Therefore, only the software DDA method is discussed in this text.

5-4 SOFTWARE DDA INTERPOLATOR

The software DDA method is an interpolation procedure which is based on the operation of hardware DDA interpolators. The DDA interpolator requires successive addition operations in order to create new interpolated points and is therefore ideally suited to assembly language simulation. The rate of iterations in the hardware interpolator is controlled by a high-frequency clock. A simple calculation in Example 5-4 has shown that this clock frequency will be in the range of hundreds kilohertz. However, since the cycle time of a minicomputer is about 1 μs, it is impossible to use such a high frequency for the interrupt pulses. In order to adapt the clock frequency to a permitted interrupt rate, the software interpolator requires some modification, essentially in the definition of register length [10].

The ratio between the machine tool feedrate V (in BLUs per second) and the input frequency to the interpolator f_0 can be determined in linear motions by combining Eqs. (5-24) and (5-26)

$$\frac{V_\ell}{f_0} = \frac{L}{2^n} \tag{5-43}$$

Similarly, for circular motions this ratio is calculated from Eqs. (5-40) and (5-41):

$$\frac{V_\ell}{f_0} = \frac{R}{2^n} \tag{5-44}$$

The efficiency of the hardware interpolator, which is defined by the ratio $L/2^n$ (or $R/2^n$) might be very low. This number represents the ratio between the actual distance of motion (L) and the maximum allowable one ($2^n - 1$). An incremental motion of a relatively small distance with a high feedrate prescribes the requirement of the high frequency f_0. Hence, the frequency f_0 can be decreased by increasing the efficiency of the interpolator. In software this can be easily done by using registers with variable lengths rather than fixed ones as in hardware interpolators. Moreover, in software the maximum allowable distance is not necessarily represented in powers of 2 and can be any number S, i.e., $S = 2^n$ is substituted in Eqs. (5-43) and (5-44). The maximum efficiency is obtained by applying a variable S, which is calculated at the beginning of each segment as follows:

For linear motion: $S = L$

For circular motion: $S = R$

L is given in Eq. (5-22) and R in Eq. (5-37). By using the variable S, Eqs. (5-43) and (5-44) reduce to

$$V_\ell = f_0 \tag{5-45}$$

which means that DDA 3 must provide an iteration rate which is equal to the velocity along the path in BLUs per second. According to Eq. (5-11), the output of this DDA is

$$f_0 = \frac{\text{FRN}}{S_0} f \tag{5-46}$$

where S_0 is the maximum allowable feedrate number code, and f is the constant frequency of the external clock, which in a software DDA becomes the source of the interrupt pulse frequency. Equations (5-45) and (5-46) show that the programmed FRN code in CNC systems must be proportional to the required feedrate, or velocity along the path (see Sec. 3-2.1).

Example 5-6 The maximum permissible feedrate in a CNC system is 3600 mm/min, and the system resolution is BLU = 0.01 mm. If the FRN is programmed directly in millimeters per minute, calculate the constant S_0 and the frequency of the interrupt clock pulse.

SOLUTION

$$S_0 = 3600$$

$$f = \frac{V_{max}}{BLU} = \frac{3600}{0.01 \times 60} = 6000 \text{ pps}$$

An interrupt frequency of 6000 pps allows time intervals of 167 μs for the execution of one iteration of the CNC control program in which the interpolator is contained.

Figure 5-13 shows the flowcharts of the feedrate routine, the linear interpolator, and the circular interpolator. Simpler DDAs with $\Delta p = 0$ are used in the simulation of the feedrate DDA and the linear interpolator, while full software DDAs are used for the circular interpolator. Note that S_0 represents in a hardware DDA the maximum content of the p register of the feedrate DDA 3. In the linear interpolator the initial conditions are

$$p_1 = a \qquad p_2 = b$$

and L is calculated according to Eq. (5-22). In the circular interpolator the initial conditions are given in Eq. (5-33):

$$p_1 = R \cos \omega t_0 \qquad p_2 = R \sin \omega t_0$$

and the radius R is calculated from Eq. (5-37).

5-5 REFERENCE-WORD CNC INTERPOLATORS

CNC systems for manufacturing utilize either the reference-pulse or the sampled-data technique. The interpolators in reference-pulse systems are based upon an iterative technique controlled by an interrupt clock. At each interrupt, a single iteration of the interpolation routine is executed, which in turn can provide an output pulse that advances the corresponding machine axis by one BLU. Therefore, the maximum attainable feedrate, or axis velocity, in BLUs per second is inversely proportional to the execution time of a single iteration. The appropriate reference-pulse interpolation technique for machine tool systems is the software DDA, since uniformity of the feedrate along a circular path is attainable only with this method. Using a PDP-11 as the CNC controller, an interpolation iteration time of 52 μs has been obtained [12], which allows a maximum velocity of 19,230 BLUs per second. A BLU might be on the order of 10 μm in a typical machine tool system, resulting in a maximum feedrate of about 11 m/min, which is adequate for machining applications. However, modern CNC systems employ a single axial drive motor for both idle motions and machining, and in these systems a velocity of 11 m/min is considered to be too low for idle motion.

In contrast, in sampled-data systems (see Fig. 7-4) the maximum velocity is not limited by the computer. The control program contains both linear and circular interpolation subroutines which function in an on-line mode. Although linear interpolators

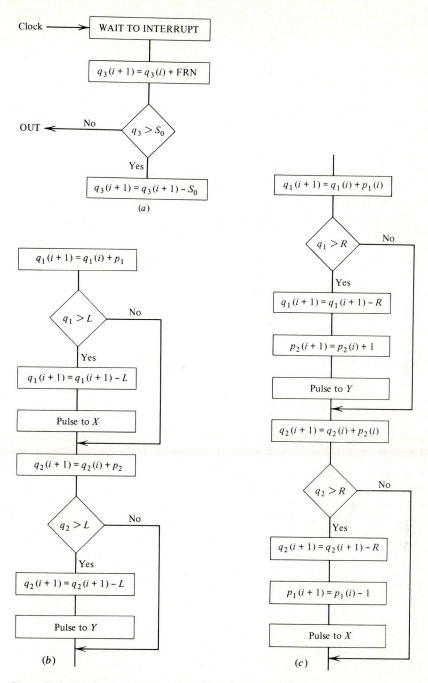

Figure 5-13 Flowcharts of a software DDA interpolator. (*a*) Feedrate subroutine; (*b*) linear interpolator; (*c*) circular interpolator.

can be easily written, even in assembly language, the difficulty arises in programming the circular interpolators which require on-line solution of a second-order equation. In sampled-data systems the circular interpolation is based upon approximating a circle with straight line segments. For each segment the interpolator generates a reference word proportional to the local axial velocities, which are transmitted to the corresponding software loop comparators of the control axes. This section introduces reference-word circular interpolator techinques which can be applied to manufacturing systems permitting the simultaneous operation of two axes of motion.

5-5.1 The Concept of Reference-Word Interpolators

In circular interpolation, the simultaneous motion of two axes generates a circular arc at a constant tangential velocity, or feedrate, V. The axial velocities satisfy the following equations:

$$V_x(t) = V \sin \theta(t)$$

$$V_y(t) = V \cos \theta(t)$$

(5-47)

where

$$\theta(t) = \frac{Vt}{R}$$

and R is the radius of the circular arc.

The velocity components V_x and V_y are computed by the circular interpolator and are supplied as reference inputs to the computer closed loops. The circle generated in this case is actually comprised of straight line segments. At the beginning of each segment the references are supplied by the interpolator and the end of the segment is located with the aid of a feedback signal. Increasing the number of these segments improves the accuracy of the generated circle but increases the number of iterations, thus requiring more computer time. The optimal number of segments is the smallest one which maintains the path error within the required limit of one BLU.

Each iteration of the algorithm corresponds to an angle α, as is illustrated in Fig. 5-14. The choice of the angle α depends on the interpolation method. All of these methods employ the difference equation:

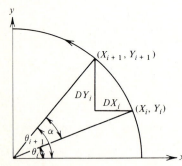

Figure 5-14 Two successive points on a circular arc.

$$\cos \theta(i + 1) = A \cos \theta(i) - B \sin \theta(i)$$
$$\sin \theta(i + 1) = A \sin \theta(i) + B \cos \theta(i)$$
$$(5\text{-}48)$$

where the coefficients A and B are given by

$$A = \cos \alpha \qquad B = \sin \alpha \qquad\qquad (5\text{-}49a)$$

and
$$\theta (i + 1) = \theta (i) + \alpha \qquad\qquad (5\text{-}49b)$$

The corresponding segment is terminated at the point $X (i + 1)$, $Y (i + 1)$, which is approximated by

$$X(i + 1) = R (i) \cos \theta(i + 1)$$
$$Y(i + 1) = R (i) \sin \theta(i + 1)$$
$$(5\text{-}50)$$

Substituting Eq. (5-48) into Eq. (5-50) yields

$$X(i + 1) = AX(i) - BY(i)$$
$$Y(i + 1) = AY(i) + BX(i)$$
$$(5\text{-}51)$$

Equation (5-51) is the basic relationship which permits the calculation of a successive point based on the present one.

The main differences between the various interpolation methods are in the determination of α and in the approximation of the coefficients A and B in Eq. (5-49a). Once the angle α is chosen, the interpolator routine proceeds, for each iteration, as follows:

1. At each point $X(i)$, $Y(i)$ the interpolator calculates the coordinates of the successive point $X(i + 1)$, $Y(i + 1)$ according to Eq. (5-51). The segment lengths are

$$DX(i) = X(i + 1) - X(i) = (A - 1)X(i) - BY(i)$$
$$DY(i) = Y(i + 1) - Y(i) = (A - 1)Y(i) + BX(i)$$
$$(5\text{-}52)$$

and the corresponding velocities are

$$V_x(i) = \frac{V \, DX(i)}{DS(i)}$$
$$(5\text{-}53)$$
$$V_y(i) = \frac{V \, DY(i)}{DS(i)}$$

where
$$DS(i) = \sqrt{DX^2(i) + DY^2(i)}$$

2. The values obtained from Eq. (5-52) are the incremental positions, and those of Eq. (5-53) are the velocities, or the reference words, for the present segment. These values are supplied to the software control loops.
3. Upon completion of the segment, the routine increments the coordinates $X(i)$ and $Y(i)$.

Since the angle α is relatively small, the chord length DS can be approximated by its arc length $R\alpha$, and the calculation of the velocities V_x and V_y can be simplified to

$$V_x(i) = K \, DX(i)$$
$$V_y(i) = K \, DY(i)$$
$$(5\text{-}54)$$

where $K = V/R\alpha$. The parameter K is a constant, which is calculated only once for each circle, and consequently the reference words are actually proportional to the segment lengths.

Approximation of a circle by straight line segments causes two types of errors, as illustrated in Fig. 5-15:

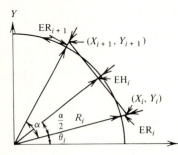

Figure 5-15 Errors in a circle generated with the reference-word interpolator.

1. Radial error ER, due to the truncation effects:

$$ER(i) = R(i) - R = \sqrt{X^2(i) + Y^2(i)} - R \qquad (5\text{-}55)$$

where R is the required radius.

2. Chord height error EH defined by

$$EH(i) = R - R(i)\cos\frac{\alpha}{2} \qquad (5\text{-}56)$$

Equation (5-55) is explicitly obtained using the approximated values of A and B, which causes the truncation error in the radius. Since the algorithms use an iterative technique, this error is accumulated with time. The effect of the truncation error ER at the ith iteration can be approximated by [14]

$$ER(i) = i(C - 1)R \qquad (5\text{-}57)$$

where

$$C = \sqrt{A^2 + B^2}$$

Equation (5-57) shows how the truncation error increases with the number of iterations. The largest circular arc produced by a single instruction in CNC systems is a quarter of a circle. The compatible number of iterations is

$$N = \frac{\pi}{2\alpha} \qquad (5\text{-}58)$$

Upon completing this number of steps the error ER reaches its maximum value:

$$ER_{max} = \frac{\pi}{2\alpha}(C - 1)R \qquad (5\text{-}59)$$

By contrast the error EH is not cumulative. Using the formula

$$\cos\frac{\alpha}{2} = \sqrt{\frac{1 + \cos\alpha}{2}} = \sqrt{\frac{1 + A}{2}} \qquad (5\text{-}60)$$

in Eq. (5-56) yields

$$EH(i) = R - R(i)\sqrt{\frac{1 + A}{2}} \qquad (5\text{-}61)$$

The angle α is selected so that either the error ER or EH reaches a maximum value of 1 BLU. In addition to the angle α, the interpolator is given the coordinates of the initial point $(X(0), Y(0))$ and the required feedrate V. These parameters are obtained from the part program for each circular arc.

Several interpolators applying the reference-word method have been presented in the literature [2, 3, 18]. It has been shown [14] that if a floating-point unit (FPU) is available in the computer, a method denoted as the *improved Tustin method (ITM)* is the favorite in terms of the required number of iterations to produce a circular arc.

The basic *Tustin method* is now discussed, and the ITM will be introduced later.

5-5.2 Tustin Method

The Tustin method [3] is based on an approximate relationship between the derivative operator s and the discrete variable z:

$$s = \frac{2}{T}\left(\frac{z - 1}{z + 1}\right) \qquad (5\text{-}62)$$

Using the Tustin approximation yields the following relations for $\cos \alpha$ and $\sin \alpha$, or A and B:

$$A = \frac{1 - (\alpha/2)^2}{1 + (\alpha/2)^2} \qquad B = \frac{\alpha}{1 + (\alpha/2)^2} \qquad (5\text{-}63)$$

These approximations can be used to generate circular interpolation for CNC systems. Substituting these values of A and B into Eqs. (5-52) and (5-54) gives the basic equations of the proposed algorithm:

$$DX(i) = -\frac{1}{1 + (\alpha/2)^2}\left[\frac{\alpha^2}{2}X(i) + \alpha Y(i)\right]$$

$$(5\text{-}64)$$

$$DY(i) = \frac{1}{1 + (\alpha/2)^2}\left[-\frac{\alpha^2}{2}Y(i) + \alpha X(i)\right]$$

$$V_x(i) = -\frac{V}{R[1 + (\alpha/2)^2]}\left[\frac{\alpha}{2}X(i) + Y(i)\right]$$

$$(5\text{-}65)$$

$$V_y(i) = \frac{V}{R[1 + (\alpha/2)^2]}\left[-\frac{\alpha}{2}Y(i) + X(i)\right]$$

The truncation error is found by substituting Eq. (5-63) into Eq. (5-57) which shows that

$$ER(i) = 0 \qquad (5\text{-}66)$$

Since the error ER is identically zero, the angle α is prescribed by the error EH given in Eq. (5-61). For $R(i) = R$

$$EH = R - \frac{R}{\sqrt{1 + (\alpha/2)^2}} \qquad (5\text{-}67)$$

which can be approximated for small α by

$$EH = \frac{\alpha^2}{\alpha^2 + 8} R \qquad (5\text{-}68)$$

The angle α is derived by equating the error EH to one unit:

$$\alpha = \sqrt{\frac{8}{R - 1}} \simeq \sqrt{\frac{8}{R}} \qquad (5\text{-}69)$$

5-5.3 Improved Tustin Method

The Tustin method yields an ER of zero. From a practical point of view in manu-facturing systems we choose the maximum error equal to 1 BLU. Therefore the method can be improved by increasing the angle α. From Fig. 5-16 the angle α is evaluated:

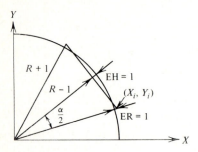

Figure 5-16 Circular arc generated by the improved Tustin method.

$$\cos \frac{\alpha}{2} = \frac{R - 1}{R + 1} \qquad (5\text{-}70)$$

Using Eq. (5-60)

$$\frac{R + 1}{R - 1} = \sqrt{\frac{2}{1 + A}} = \sqrt{1 + \left(\frac{\alpha}{2}\right)^2} \simeq 1 + \frac{\alpha^2}{8} \qquad (5\text{-}71)$$

so that the angle α is

$$\alpha = \sqrt{\frac{16}{R - 1}} \simeq \frac{4}{\sqrt{R}} \qquad (5\text{-}72)$$

Therefore, compared to Eq. (5-69), the angle α is increased by a factor of $\sqrt{2}$. The number of iterations to produce a quarter of a circle, according to Eq. (5-58), is

$$N = \frac{\pi}{8} \sqrt{R} \qquad (5\text{-}73)$$

Example 5-7 Compare the number of iterations required to produce a quarter of a circle with $R = 10,000$ BLUs by the Tustin method, the ITM, and the software DDA method.

SOLUTION According to Eq. (5-69) the angle α with the Tustin method is $\alpha = \sqrt{8/10,000}$. The number of iterations is $N = \pi/2\alpha = 56$ (N is rounded to the next integer). The number of iterations with the ITM is obtained from Eq. (5-73): $N = 40$. The number of iterations with the software DDA is equal to the length of the arc in BLUs [10]: $N = \pi R/2 = 15,708$.

BIBLIOGRAPHY

1. Angelov, A. S., A. D. Mihailov, and G. N. Nachev: Simple Algorithm for Curvilinear Interpolation, *Proc. 11th Ann. Meet. Tech. Conf. NC Soc.*, pp. 338–348, April, 1974.
2. Bergren, C.: A Simple Algorithm for Circular Interpolation, *Contr. Eng.*, pp. 57–59, September, 1971.
3. Cadzow, J. A., and H. R. Martens: "Discrete-Time and Computer Control Systems," chap. 9, Prentice-Hall, Inc., Englewood Cliffs, New Jersey, 1970.
4. Danielson, P. E.: Incremental Curve Generation, *IEEE Trans. Comp.*, vol. C-19, no. 9, pp. 783–793, September, 1970.
5. Gorman, J. E., and J. Raamot: Integer Arithmetic Technique for Digital Control Computers, *Comp. Des.*, pp. 51–57, July, 1970.
6. Jordan, B. W., W. J. Lenon, and B. D. Holm: An Improved Algorithm for Generation of Nonparametric Curves, *IEEE Trans. Comp.*, vol. C-22, no. 12, pp. 1052–1060, December, 1973.
7. Kaiwa, T., and S. Inaba: Latest Japanese Numerical Control Features, *Contr. Eng.*, pp. 88–91, October, 1961.
8. Koren, Y., A. Shani, and J. Ben-Uri: Numerical Control of a Lathe, *IEEE Trans. Ind. Gen. Appl.*, vol. 6, no. 2, pp. 175–179, March, 1970.
9. ———: Design Concepts of a CNC System, *Proc. IEEE Ind. Appl. Soc.*, Annual Meeting, Atlanta, pp. 275–282, October, 1975.
10. ———: Interpolator for a CNC System, *IEEE Trans. Comp.*, vol. C-25, no. 1, pp. 32–37, January, 1976.
11. ———: Design of Computer Control for Manufacturing Systems, *Trans. ASME, J. Eng. Ind.*, vol. 101, no. 3, pp.326–332, August, 1979.
12. ———, and O. Masory: Reference-Pulses Circular Interpolators for CNC Systems, *Trans. ASME, J. Eng. Ind.*, vol. 103, no. 1, pp. 131–136, February, 1981.
13. Masory, O., and Y. Koren: The Direct Search Method in CNC Interpolators, *ASME-78-WA/PROD-40*, December, 1978.
14. ———, and Y. Koren: Reference-Word Circular Interpolators for CNC Systems, *Trans. ASME, J. Eng. Ind.*, vol. 104, November, 1982.
15. Mayrov, F. V.: "Electronic Digital Integrating Computers—Digital Differential Analyzers," Iliffe Books, London, England, 1964.
16. McGhee, R. B., and R. N. Nilsen: The Extended Resolution Digital Differential Analyzer: A New Computing Structure for Solving Differential Equations, *IEEE Trans. Comp.*, vol. C-19, pp. 783–793, September, 1970.
17. Milner, D. A.: Some Aspects of Computer Numerical Control with Reference to Interpolation, *Trans. ASME, J. Eng. Ind.*, pp. 883–889, August, 1976.
18. Musse, J., M. Veron, F. Lepage, and J. Drapier: High Performance Tool Path Interpolator for CNC System, *CIRP Mfg. Sys.*, vol. 7, no. 1, 1978.
19. Sizer, T. R.: "The Digital Differential Analyzer," Chapman & Hall, London, England, 1968.

PROBLEMS

5-1 A DDA contains 3-bit registers which are initially set to $p = 6$ and $q = 0$. Calculate the output Δz at the first 8 iterations, assuming that each iteration is executed in 1 ms. Draw the accumulated output Δz versus time and measure the maximum error between the desired line $z = 750t$ and the actual one.

5-2 A DDA contains 8-bit registers, the value of its p register is constant $p = 150$, and the clock frequency is 10,240 pps. Calculate the output frequency of the DDA.

5-3 The Δz output in Fig. 5-5 is connected through a frequency divider by 2 to the $-\Delta p$ input. (In practice, a trigger flip-flop is fed by the Δz pulses, and its output is connected to the $-\Delta p$ input).

(a) What is the differential equation which is solved with this circuit?

(b) If the DDA contains 4-bit registers, the clock frequency is 16 pps. and $p_0 = 15$, determine the expression $p(t)$.

5-4 The Δz output in Fig. 5-5 is connected to the $+\Delta p$ input (instead of $-\Delta p$). The initial condition is $p = 1$ and the register's length is $n = 3$. The iteration time is 1 ms.

(a) Calculate the iterative values of p as a function of time ($p < 8$). Draw the diagram $p = f(t)$.

(b) Derive the function $p = f(t)$.

5-5 Design a decelerator circuit which consists of a DDA with 9-bit registers and a frequency divider by k. The circuit decelerates a frequency of 500 pps to approximate 95 pps during 10 s in an exponentially decaying rate. The clock frequency is 512 pps.

(a) Find the initial value of the p register.

(b) Determine the exponent α in Eq. (5-13).

(c) Calculate the division factor k.

5-6 How many different slopes can be achieved on a NC milling machine which uses a DDA linear interpolator with registers of $n = 3$-bit length?

5-7 Given a DDA circular interpolator with $n = 5$, and initial conditions $i = 31, j = 0$. Calculate the values of the p and q registers at each interpolation step. Arrange the results in a table similar to Table 5-4. Draw the obtained circular arc. Find the maximum error in the obtained circle.

5-8 The XY table of a drilling machine has stepping motor drives. Each pulse causes a motion of BLU $= 0.01$ mm. The required velocity of each axis is 2.8 mm/s. The NC system contains a pulse generator which supplies a constant frequency of 640 pps.

A DDA integrator which acts as a frequency divider is used as an interface between the pulse generator and the stepping motor. What is the minimum number of binary stages in the DDA, and what is the contents of its p register? Draw a block diagram.

5-9 In order to accelerate gradually the stepping motor in the previous problem, it is suggested to start with a zero value in the p register and add 1 every 1/160 s (each fourth iteration of the DDA) until the required value of p is reached. What is the acceleration distance in millimeters?

5-10 A linear interpolator which consists of three DDA integrators (see Fig. 5-12) is loaded by data written in the inverse-time code and drives two stepping motors. DDA 3 contains 8-bit registers, and 1 and 2 are of 15 bits each

(a) Calculate the clock frequency f.

(b) For an incremental position of $\Delta X = 4000$ BLU, $\Delta Y = 3000$ BLU, and feedrate of 60,000 BLU/min, what is the value of the p register of DDA 3 and what are the frequencies f_0, f_x, and f_y? Check whether the required feedrate is obtained.

5-11 The execution time of a software DDA interpolation routine is 52 μs. If the system resolution is BLU $= 0.01$ mm, calculate the maximum axial feedrate in meters per minute.

5-12 Write a computer program for a software DDA circular interpolator. Use a high-level language such as BASIC or FORTRAN. Try your program with the data given in Example 5-5 and compare the results with Table 5-4.

5-13 A CNC system applies the ITM reference-word interpolator. Assuming that each iteration is executed in 1 ms, what is the maximum attainable velocity around a circle of $R = 225$ mm, if the system BLU is 0.01 mm?

CONTROL LOOPS OF NC SYSTEMS

This chapter deals with the analysis and design of open- and closed-loop controllers for manufacturing systems. Open-loop controllers have no access to any real-time information about the system performance and therefore cannot counteract disturbances appearing during the operation. They can be utilized in point-to-point systems, where the loading torque on the axial motors is almost constant.

Closed-loop controllers are much more powerful than their open-loop counterparts. They measure the state of the system during operation, and therefore they can reduce the effects of load disturbances and compensate in real-time for parameter variations.

6-1 INTRODUCTION

The control loops of NC systems are designed to perform a specific task: to control the position and velocity of the machine tool axes. In NC systems each axis is separately driven and should follow the command signal produced by the interpolator. The system design starts by selecting the type of control: open-loop or closed-loop, a decision which depends on the required specifications of the NC system and economy. Open-loop controls use stepping motors as the drive devices of the machine table, as shown in Fig. 6-1. The drive units of the stepping motors are directly fed by the interpolator

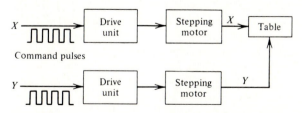

Figure 6-1 Open-loop control system for a two-axis machine.

output pulses. The selection of the appropriate motor depends on the maximum torque, required velocity, and step size in the system. Stepping motors can be implemented on small-sized point-to-point systems in which the load torque is small and constant.

In the design of closed-loop systems the engineer might first choose the appropriate drive device (hydraulic or dc motor) and the feedback element (encoder, resolver, or inductosyn). The actual design depends on this selection as well as on the functional objective of the system: point-to-point or contouring. In point-to-point systems the workpiece is mounted on the machine table, which moves with respect to the tool. In these systems, the path of the table with respect to the cutting tool and its velocity while moving from one point to the next one are without any significance. Therefore, point-to-point systems require only control of the final position of the table at each motion. In order to save machine time, the table travels between the points at high velocities. Since the drive system and the machine table usually have appreciable inertia, the motion must be gradually decelerated before reaching the point to be drilled. Appropriate deceleration circuits are included in point-to-point control systems.

In contouring systems the tool is cutting while the machine axes are moving. The contour of the part is determined by the ratio between the axial velocities together with the position of the tool at the end of each segment. Therefore contouring systems must contain digital counters which check the end position of each segment in addition to *position control loops*.† The control in contouring systems operates in closed loops, which compare the command pulses from the interpolator with the feedback signal from the encoder or the resolver. In this text only control loops which use encoders are described. The digital circuits associated with a resolver feedback are more complicated, and their design concepts can be found in the literature [4, 9, 12].

The most sophisticated design applies to the closed-loop control of contouring systems. In the design of these loops, the transfer function of each element must first be determined; the system is then set up in block diagram form, and finally the loop gain is established based on performance analysis. The transfer function of each element is based upon its mathematical model. In establishing the mathematical model the engineer is faced with a compromise between *accuracy* and *complexity* on one hand, and *approximation* and *simplicity* on the other. In this chapter we shall discuss simple models, with which the principles of design can be readily illustrated.

6-2 CONTROL OF POINT-TO-POINT SYSTEMS

In point-to-point systems each axis is driven separately at the maximum allowable velocity. This velocity depends on the drive type and on the mechanical structure of the particular manufacturing system. In order to avoid large overshoots the velocity is decelerated before the target point in which the tool starts to operate (e.g., drill). Since

---

† In control literature, loops with similar structure are referred to as *velocity control loops* and in manufacturing literature they are referred to as *position control loops*, a term which is a misnomer in describing the task of these loops. Nevertheless, in order to permit communication with manufacturing engineers the term position control loops is used in this text.

the path between the points is insignificant, the deceleration is accomplished in each axis separately.

The backlash in the axial gear and between the leadscrew and the table's nut affects the accuracy of slide positioning in both open- and closed-loop systems. The position feedback transducer in closed-loop systems is mounted on the leadscrew, and therefore even in this case the mechanical linkage between the leadscrew and the table functions in open loop. In point-to-point systems the backlash effect can be eliminated by using a *backlash take-up circuit*. This circuit causes the approach to the target position to always take place from the same direction. If the table approaches the target point in one direction, then the deceleration procedure takes place and the table stops without overshoot. However, when approaching the target point in the other direction, the table is not decelerated and will overrun the target; only then is the table decelerated and reversed to approach the target position in the first direction. This method cannot be used in contouring systems, in which overshooting the target point causes overcut of the material and subsequent errors in the produced part.

Point-to-point systems can be designed to use incremental or absolute programming. The main control elements in incremental point-to-point systems are position counters; each axis of motion is equipped with a down-counter. The counter is loaded to the required axial incremental position in BLUs, which is given by the corresponding dimension word of the part program and subsequently decremented by pulses, where each pulse indicates an axial motion of one BLU. By this method the contents of the counter continuously represent the distance to the target point, which can be decoded and used to trigger deceleration circuits.

6-2.1 Incremental Open-Loop Control

Open-loop point-to-point controls utilize the incremental method which was introduced in Chap. 1. A block diagram of a typical control of a single axis is shown in Fig. 6-2.

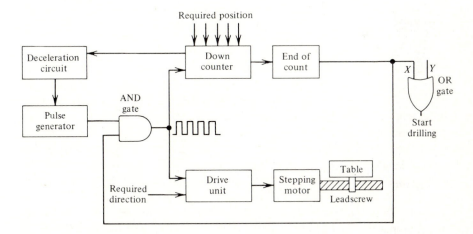

Figure 6-2 Incremental open-loop control for PTP system.

The leadscrew of the table is driven by a stepping motor, where each step of the motor advances the table by 1 BLU. At the beginning of each programming block the tool is located above the workpiece and the down-counter is loaded to the required incremental axial distance in BLUs. A pulse generator supplies pulses which simultaneously drive the stepping motor and decrement the contents of the counter. The counter reaches a zero value when the motor has stepped through the required number of steps. The zero position, which is decoded by the end-of-count circuit, changes the output of this circuit from 1 to 0 logic level. As a consequence the AND gate blocks the flow of pulses from the pulse generator, which, in turn, stops the motor and the counting.

When both the X and the Y counters are at zero, the tool is located in the required position and is ready to operate. The tool approaches the workpiece, drills, and retracts with the aid of a sequential limit switch control which is adjusted on the machine for each machining job. The tool retraction provides a signal to the tape reader to read the next block of the part program and the operation continues.

6-2.2 Incremental Closed-Loop Control

The main disadvantage of the open-loop control is the lack of information regarding the actual motion of the table leadscrews. This can be remedied by mounting a feedback transducer, such as an encoder, to the other side of the leadscrew, as shown in Fig. 6-3.

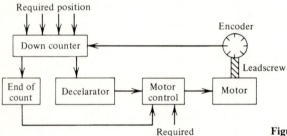

Figure 6-3 Block diagram of closed-loop incremental PTP system.

The down-counter is loaded to the required incremental position, which resets the end-of-count circuit and allows the motor to rotate. The down-counter is fed by the encoder pulses which represent the actual motion, rather than by the command pulses which are produced by the pulse generator in the open-loop control. The motor rotates as long as the contents of the counter is not zero; namely, as long as the required position has not been reached. Upon reaching the desired position, the zero contents of the counter activate the end-of-count circuit, which in turn stops the motor. The decelerator circuit slows down the motor before the target point in order to avoid overshoot.

A typical three-stage deceleration diagram of one axis of the table is given in Fig. 6-4. The table moves at rapid velocity V until reaching a distance $L1$ from the target point, where the table is instructed to move at smaller velocity $V1$. After a time delay,

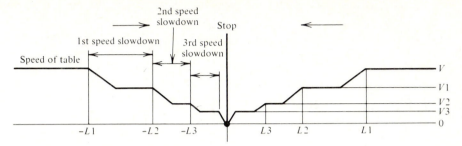

Figure 6-4 Deceleration diagram in PTP system.

which depends on the system inertia, the table moves at the new velocity $V1$ until reaching distance of $L2$ units from the target point, where again the velocity is reduced to $V2$. When the table is at a distance of $L3$ units before the target point, the velocity is reduced once more and the table "creeps" toward the final point at very low velocity $V3$, and subsequently stops.

> **Example 6-1** The deceleration distances in a particular NC point-to-point system are $L1 = 10$ mm, $L2 = 1$ mm, $L3 = 0.1$ mm, before the target point. The corresponding velocities are 40, 4, and 0.4 mm/s. Calculate the deceleration time (assume that transients are negligible).
>
> SOLUTION
>
> $$t = \frac{L1 - L2}{V1} + \frac{L2 - L3}{V2} + \frac{L3}{V3} = \frac{9}{40} + \frac{0.9}{4} + \frac{0.1}{0.4} = 0.7 \text{ s}$$

Although a stepping motor can theoretically be used in closed-loop systems, in practice only dc motors are utilized. A schematic diagram of one axis driven by a dc motor is shown in Fig. 6-5. The required incremental position of the axis is given by the corresponding dimension word of the part program and is loaded into the position down-counter. As a consequence the motor starts to rotate and the contents of the counter is gradually reduced by the pulses from the encoder. The motor control unit consists of three relays, which are excited by corresponding decoder circuits and provide the required operating voltage to the motor. At the beginning of the motion none of the relays is active, and a voltage U is supplied to the motor terminals, resulting in a corresponding speed V. When the motor reaches a distance of $L1$ BLUs before the target point, the contents of the counter is also $L1$, and the corresponding decoder becomes active, which consequently excites the relay $R1$. As a result a smaller voltage $U1$ is supplied to the motor; this state is shown in Fig. 6-5. Similarly when a distance $L2$ is reached, the relay $R2$ is excited, and a very small voltage $U2$ is supplied. The motor "creeps" to the final position, which is detected by the end-of-count circuit (or a zero decoder). The latter excites the relay $R3$, which grounds the motor terminals and simultaneously activates a mechanical brake which completely stops the movement of the table.

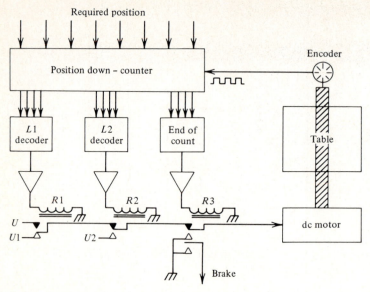

Figure 6-5 Incremental closed-loop PTP system of one axis.

6-2.3 Absolute Closed-Loop Circuit

The absolute positioning system utilizes absolute programming in conjunction with an incremental feedback device. The heart of the control loop consists of three units, which are shown in Fig. 6-6. The *position register* is fed by two alternative sequences

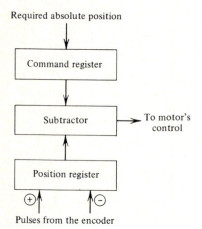

Figure 6-6 The structure of the loop comparator in an absolute closed-loop control system.

of pulses from the incremental encoder, one for each direction of motion. Its contents are increased for a rightward movement of the corresponding axis and are reduced for a leftward motion, thus indicating the actual absolute position of the axis. The *command register* is loaded with the required absolute position of the axis, which is given by the corresponding dimension word in the part program. Each time a new block is read, the command register is reloaded with a new positioning command. The *subtractor unit* indicates the instantaneous actual difference between the required and actual position, which is the distance to the target point. The subtractor output is the position error of the loop, which is fed through a deceleration circuit to the motor. Comparing this structure with the incremental one, we see that the three units of Fig. 6-6 replace the down-counter, which was the heart of the incremental loop.

Example 6-2 The part program in an absolute NC system contains the following successive X dimension words: 0, 1000, 5000, 3000. Determine the contents of the position and command registers at the beginning of each block.

SOLUTION

Command	Position	Subtractor
0	0	0
1000	0	1000
5000	1000	4000
3000	5000	− 2000

The output of the subtractor circuit always indicates the required incremental position to the next point.

6-3 CONTROL LOOPS IN CONTOURING SYSTEMS

Contouring systems (also called continuous-path systems) are used with lathes, milling machines, grinders, jig borers, laser-beam cutters, welders, certain types of robots, and other manufacturing systems. In all these systems the machine axes are separately controlled and should follow reference signals (in form of pulses) generated by the interpolator. The interpolator coordinates the motion along the system axes by supplying the corresponding sequences of reference pulses for each axis of motion, where each pulse generates a motion of 1 BLU of axis travel. Unlike point-to-point systems in which the velocity while traveling from one point to the next has no particular significance, in contouring systems accurate velocity control of each axis is of extreme importance, since cutting takes place during the motion of the axes. Any deviation in an axial velocity causes an error in the shape of the part. Therefore, a contouring system must control both the position and the velocity of each axis of motion, which results in a more complicated system.

6-3.1 Principle of Operation

The control loops of contouring systems are usually of the closed-loop type as shown in Fig. 6-7. They use two feedback devices: a tachometer which is included in the drive

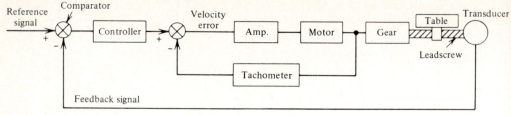

Figure 6-7 Control loop of contouring system.

unit (see Fig. 4-11) and a second feedback transducer which is capable of measuring both position and velocity (such as an encoder, resolver, or inductosyn) and therefore can be applied in position control loops. In encoder-based systems the encoder is mounted on the leadscrew and emits voltage pulses; each pulse indicates a motion of 1 BLU of axis travel. Therefore, the number of pulses represents position and the encoder pulse frequency is proportional to the axis velocity. In resolver-based systems, the rotor of the resolver is directly coupled to the machine leadscrew. The output voltage of the resolver when rotating at constant speed is given by Eq. (4-22), which is repeated here

$$V_a = V \sin [(\omega + \omega_0)t + \phi_0] \qquad (6\text{-}1)$$

where $\omega = 2\pi f$

$\quad f$ = excitation voltage frequency

$\quad \omega_0$ = angular velocity of the rotor

$\quad \phi_0$ = cumulative angle during the transient state until reaching the steady state

Again, the term $(\omega_0 t + \phi_0)$ represents position, and the frequency ω_0 is proportional to the axis velocity.

The main element in closed-loop controls is the comparator, which compares the reference and feedback signals, as shown in Fig. 6-7. In a resolver-based system this element is denoted as a *phase comparator,* or discriminator. The reference signal to the phase comparator is a square wave with a frequency of $(f + f_1)$, where f_1 is the required leadscrew velocity in revolutions per second. The frequency f_1 is generated by the interpolator and a series of variable-count counters (VCC). The phase comparator subtracts the feedback frequency $(f + f_0)$ from the reference frequency $(f + f_1)$ and determines the phase difference

$$\Delta\phi = \int_0^t (f_1 - f_0) \, dt$$

which is used to drive the motor. Detailed descriptions of the operation of the VCC and the resolver-based control loop can be found in the literature [4, 9, 12].

The principle of operation of the encoder-based control loop is simpler, but its block diagram representation and the analysis are similar to those in a resolver-based system. An encoder-based loop consisting of an up-down counter, a DAC, and a drive unit is shown in Fig. 6-8. The drive unit contains an internal loop consisting of

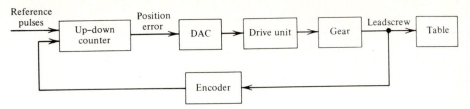

Figure 6-8 Encoder-based control loop for contouring system.

the power amplifier, dc motor, and the tachometer as a second feedback device, as shown in Fig. 6-7.

The comparator element in Fig. 6-8 is the up-down counter which is fed by two sequences of pulses: reference pulses from the interpolator and feedback pulses generated by the encoder. The counter produces a number representing the instantaneous *position error* in pulse units. This number is converted by the DAC to a voltage which is amplified and applied to the motor. The motor rotates in the direction that reduces the error. If a constant input frequency is applied, the encoder frequency at the steady state is equal to the reference frequency, except for a finite pulse and phase difference which is necessary to generate the corrective error voltage to rotate the motor.

Example 6-3 A straight-cut milling of an aluminum workpiece is performed along the X axis at 12 in/min. If the system resolution is BLU = 0.0001 in, what is the frequency of the reference pulses to the control loop?

SOLUTION

$$f_r = \frac{12}{60 \times 0.0001} = 2000 \text{ pps}$$

The encoder output frequency at steady state is also 2000 pps.

A typical output signal of the up-down counter at steady state for an axial motion at constant velocity is shown in Fig. 6-9. Generally, the number E in the counter at steady state is not constant, and it varies between two successive values (e.g., between three and four pulses). Each pulse that comes from the interpolator increases the number in the counter, and each pulse from the encoder decreases the count. At high input frequencies, the motor smoothes the error signal and follows the average value, but at low input frequencies the motor moves in steps.

The encoder-based loop is a digital loop which precisely controls the *velocity* of the motor (see footnote on page 144). If the motor slows down due to a higher torque demand, the feedback frequency from the encoder decreases as well, causing the pulse difference at the counter to increase. Consequently the voltage across the motor is increased as well, and the motor overcomes the excess torque demand. Similarly if the torque decreases, the voltage across the motor is temporarily reduced until the steady-state velocity is reached again. When the input frequency varies with time, the average

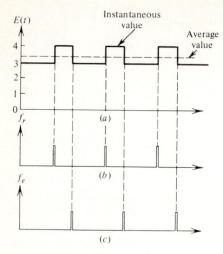

Figure 6-9 (*a*) The steady-state error in the up-down counter; (*b*) the input frequency to the counter; (*c*) the feedback frequency at steady state.

number in the counter also becomes a time variable. In this case, the error signal is not constant and depends upon the input frequency variations.

The digital control loop illustrated in Fig. 6-8 can rotate the motor in only one direction. In practice, an encoder-based loop can rotate the motor in both directions, for which an additional digital input circuit is required [8]. The input circuit precedes the counter and directs the reference and feedback pulse sequences. For one direction of rotation, the count is increased by the reference and reduced by the feedback, and vice versa for reversed rotation. The input circuit also eliminates a simultaneous appearance of pulses in both channels, which would interfere with the counting. Information about the actual direction of rotation is obtained from the encoder which feeds two sequences of square waves with 90° phase shift into a direction-sensing circuit. This provides a signal indicating the actual direction of rotation and ensures that the loop is of a negative-feedback type. By applying an appropriate bias voltage, the DAC output can be varied over a negative-to-positive voltage range to control the speed and direction of rotation of the motor. This bias together with the direction sensing and the input circuits also guarantees that the axis is locked in its position (without applying brakes) when no motion is required.

6-3.2 Mathematical Analysis

Although the control loop which has been shown in Fig. 6-8 is of a digital type, it can be analyzed by using Laplace transform techniques. The transfer function of the drive unit is given by Eq. (4-28):

$$\omega(s) = \frac{K_1 V_c(s) - K_2 T_s(s)}{1 + s\tau} \qquad (6\text{-}2)$$

where ω = motor speed
V_c = input voltage to the drive unit
T_s = static load torque on the motor

The constants in Eq. (6-2) are

$$
\begin{aligned}
K_1 &= \alpha K_a K_m, \ \text{r/(V·s)} \\
K_2 &= \frac{\alpha R K_m}{K_t}, \ \text{r/(N·m·s)} \\
\tau &= \alpha \tau_m, \ \text{s} \\
\alpha &= \frac{1}{1 + K_a K_m K_p}
\end{aligned}
\qquad (6\text{-}3)
$$

The gain terms in Eqs. (6-3) are defined as follows:

K_a = amplifier voltage gain
K_m = motor constant, r/V·s
K_t = torque constant, N·m/A
K_p = tachometer constant, V·s/r
R = armature resistance, Ω
τ_m = mechanical time constant s defined in Eq. (4-11)

There are NC systems in which the controller does not contain tachometers and internal loops. In these cases K_p becomes zero, and consequently $\alpha = 1$ is substituted in Eqs. (6-3). However, the mathematical representation given by Eq. (6-2) is still valid.

The dc motor drives the leadscrew through a gear box. The gear ratio K_g is defined as the ratio between the speed of the leadscrew ω_ℓ and the speed of the motor

$$
K_g = \frac{\omega_\ell}{\omega} \qquad (6\text{-}4)
$$

The inertia of the leadscrew and its torque should be referred to the motor shaft by Eqs. (4-15) and (4-16), respectively.

The encoder is mounted directly on the leadscrew. Its gain K_e is defined as the number of pulses produced per revolution of the encoder. This gain, however, is defined also as follows:

$$
K_e = \frac{f_e}{\omega_\ell} \qquad (6\text{-}5)
$$

where f_e is the pulse frequency transmitted by the encoder in pulses per second, and ω_ℓ is the leadscrew speed in revolutions per second.

The transfer function of the up-down counter can be determined by investigating its function. The counter has two inputs to which the reference frequency f_r and the encoder frequency f_e are fed. The counter yields a number E, which is the difference between the total number of the reference pulses N_r and the total number of the encoder feedback pulses N_e. Since each pulse represents an incremental motion of one BLU, the pulse difference E represents the position error in the loop. The pulse number difference during a time period t is calculated as follows:

$$E(t) = N_r - N_e = \int_0^t f_r \, dt - \int_0^t f_e \, dt = \int_0^t (f_r - f_e) \, dt \qquad (6\text{-}6)$$

In a block diagram representation the up-down counter is symbolized by a comparator followed by an integral controller, and in Laplace transform notation its output is given by

$$E(s) = [f_r(s) - f_e(s)]\frac{1}{s} \qquad (6\text{-}7)$$

The connection between the DAC and the up-down counter was illustrated in Fig. 4-14. The pulse number which is contained in the counter is converted to a proportional analog voltage by the DAC (see Fig. 4-15). The DAC has a conversion gain of K_c volts per pulse, which is defined as the change in its output voltage per bit (see Example 4-5 and Eqs. 4-34 and 4-37). For example, an 8-bit DAC which has an output voltage range of 10 V has a conversion gain of

$$K_c = \frac{10}{2^8} \simeq 0.04 \text{ volts per pulse}$$

The DAC output is related to the pulse number in the counter by the equation

$$V_c = K_c E \qquad (6\text{-}8)$$

The complete model of the loop is presented in Fig. 6-10. Combining Eqs. (6-2)

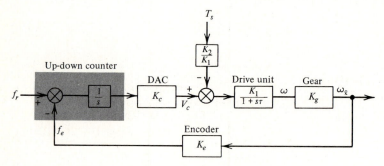

Figure 6-10 Block diagram of the encoder-based control loop.

through (6-8) gives the closed-loop response in Laplace notation:

$$\omega_\ell(s) = \frac{K_0 f_r(s) - s K_g K_2 T_s(s)}{\tau s^2 + s + K} \qquad (6\text{-}9)$$

where K is the *open-loop gain*

$$K = K_c K_1 K_g K_e \qquad (6\text{-}10)$$

and K_0 is the forward gain

$$K_0 = K_c K_1 K_g$$

From Eq. (6-9), it is seen that the closed loop behaves as a second-order system with the characteristic equation

$$s^2 + 2\zeta\omega_n s + \omega_n^2 = 0 \tag{6-11}$$

where the damping factor is

$$\zeta = \frac{1}{2\sqrt{K\tau}} \tag{6-12}$$

and the natural frequency is

$$\omega_n = \sqrt{\frac{K}{\tau}} \tag{6-13}$$

Eliminating τ from Eqs. (6-12) and (6-13) gives

$$\omega_n = 2\zeta K \tag{6-14}$$

By substituting Eq. (6-9) into Eqs. (6-5) and (6-7) the Laplace transform of the position error held in the counter becomes

$$E(s) = \frac{(1 + s\tau)f_r(s) + K_e K_g K_2 T_s(s)}{\tau s^2 + s + K} \tag{6-15}$$

For the constant reference frequency and constant load torque at the steady state, the position error held at the counter is

$$E_{ss} = \frac{f_r}{K} + \frac{K_e K_g K_2 T_s}{K} \tag{6-16}$$

It should be noted again that in the above conditions at steady state $f_e = f_r$, and the motor rotates at constant speed. E_{ss} is the difference between the total number of reference pulses and those of feedback pulses. The counter should be big enough to accommodate the number E_{ss} and overshoots in E during the transient period.

Example 6-4 Assuming that the open-loop gain is $K = 20$ s^{-1}. Calculate the required minimum number of stages in the counter to accommodate the axial velocity in Example 6-3.

SOLUTION If the load torque and overshoots are negligible, the steady-state number in the counter is

$$E_{ss} = \frac{2000}{20} = 100$$

The number of binary stages (flip-flops) n in the counter must satisfy the equation

$$2^n > E_{ss}$$

which yields $n = 7$. One bit of the counter is reserved as the sign bit (to allow reversed motion), and therefore the counter must contain at least eight binary stages.

Clearly, E_{ss} in Eq. (6-16) cannot be an integer throughout the whole speed interval. At those few speeds where it is an integer, the motor is fed by a direct voltage. Normally the counter output consists of a direct component and rectangular pulses as was shown in Fig. 6-9.

The time responses of ω_ℓ and E, for a constant input frequency and constant load torque, are found by an inverse Laplace transform of Eqs. (6-9) and (6-15)

$$\omega_\ell(t) = [1 - Q(t) \sin(\omega_d t + \phi)]\frac{f_r}{K_e}$$
$$- 2\zeta K_g K_2 T_s Q(t) \sin \omega_d t \tag{6-17}$$

$$E(t) = \frac{f_r}{K} - \frac{f_r}{\omega_n} Q(t) \sin(\omega_d t + \psi)$$
$$+ [1 - Q(t) \sin(\omega_d t + \phi)]\frac{K_e K_g K_2 T_s}{K} \tag{6-18}$$

where $\omega_d = \sqrt{1 - \zeta^2}\, \omega_n$

$$\phi = \arccos \zeta; \; \psi = 2\phi = \arctan 2\zeta\sqrt{1 - \zeta^2}/(2\zeta^2 - 1)$$
$$Q(t) = [\exp(-\zeta\omega_n t)]/\sqrt{1 - \zeta^2}$$

The time response of the leadscrew velocity ω_ℓ for various values of damping factor ζ and negligible T_s is shown in Fig. 6-11. It is seen that small damping factors

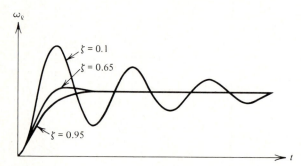

Figure 6.11 Closed-loop velocity response for various damping factors.

(e.g., $\zeta = 0.1$) cause an oscillatory response and high damping (e.g., $\zeta = 0.95$) results in a slower responding system. Therefore the damping is usually adjusted to be in the range $0.6 < \zeta < 0.8$, which is regarded as near optimal.

The actual table position $X(t)$ in BLUs is given by

$$X(t) = K_e \int_0^t \omega_\ell \, dt \tag{6-19}$$

Alternatively $X(t)$ can be calculated by subtracting the position error E from the required position, which for a linear motion yields

$$X(t) = f_r t - E(t) \tag{6-20}$$

For a negligible T_s the actual position at steady state is

$$X(t) = f_r t - \frac{f_r}{K} \tag{6-21}$$

The maximum position error for a negligible T_s occurs at

$$\omega_d t = \arctan \frac{\sqrt{1 - \zeta^2}}{\zeta} = \pi - \phi$$

which recalls an overshoot of

$$P = \frac{f_r}{2\zeta K} \exp\left[-\frac{\zeta(\pi - \phi)}{\sqrt{1 - \zeta^2}} \right] \tag{6-22}$$

above the steady state f_r/K.

The overshoot percentage of the counter versus the damping factor is plotted in Fig. 6-12. This graph can be used to calculate the maximum allowable capacity of the up-down counter.

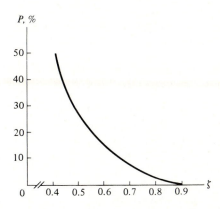

Figure 6-12 The overshoot percentage of the counter versus damping factor.

6-3.3 Design for Constant Input Frequency

Design of the digital control loop for an NC manufacturing system is generally performed for given servomotor characteristics, kinematic limitations of the machine, and precision requirements for the manufacturing process.

Selection of servomotors for NC machine tools is based upon power and torque requirements of the machine which determine the maximum motor speed ω_m and motor constant K_m. Likewise, the maximum allowable velocity for each axis of motion is dictated by kinematic considerations of the machine tool. Since each reference pulse is equivalent to the position resolution unit of the machine, this velocity limitation is equivalent to limiting the reference-pulse frequency. The maximum reference frequency in pulses per second is given by

$$f_m = \frac{V_m}{60 \times \text{BLU}} \tag{6-23}$$

where V_m is the maximum axial velocity, or feedrate, in BLUs per minute. For example, a typical system with a maximum allowable feedrate of 30 in/min and a BLU = 0.0001 in, has a maximum reference frequency of 5000 pps according to Eq. (6-23).

At the maximum reference frequency, the axis moves at its highest designed feedrate which corresponds to the maximum motor speed ω_m. The relationship between this speed and the maximum frequency is

$$f_m = K_g K_e \omega_m \tag{6-24}$$

Another parameter which is given is the leadscrew pitch (LP), given in millimeters or inches. For a single start screw the pitch is equal to the axial distance traveled per revolution of the screw. This parameter, together with the resolution unit, gives the encoder gain

$$K_e = \frac{\text{LP}}{\text{BLU}} \text{ pulses per revolution} \tag{6-25}$$

Typical values in SI units are LP = 5 mm (per revolution) and BLU = 0.01 mm (for each pulse), resulting in K_e = 500 pulses per revolution.

Since K_e, f_m, and ω_m are known, the gear ratio K_g is determined from Eq. (6-24).

The open-loop gain K is calculated from Eq. (6-12) for a given time constant τ of the motor when coupled to the machine table. Higher open-loop gains K provide a smaller position error, as seen from Eq. (6-16), but they also lower the damping factor (Eq. 6-12). Generally, K should be high for better system accuracy and faster response, but the maximum gain allowable is limited due to undesirable oscillatory response at high gains. In many servosystems it is customary to select a damping factor of $\sqrt{2}/2$, which is regarded as optimal. Designing for this damping factor, the open-loop gain is

$$K = \frac{1}{2\tau} \tag{6-26}$$

The next step is to determine the number of stages of the up-down counter and the DAC, which depends upon the maximum values of the variables. The counter must be capable of handling a motor speed of ω_m and a corresponding input frequency of f_m. According to Eq. (6-22) the overshoot for $\zeta = 0.707$ is 7 percent. This together with Eq. (6-16) gives

$$E_{\max} = \left(\frac{f_m}{K} + \frac{K_e K_g K_2 T_s}{K}\right) 1.07 \tag{6-27}$$

The number of stages n of the up-down counter, and the DAC (including one stage serving as a sign bit and indicating the direction of rotation) is derived from

$$E_{\max} \leq 2^{n-1} - 1 \tag{6-28}$$

In contouring systems, each control loop must operate linearly to maintain path accuracy. The condition in Eq. (6-28) guarantees that the counter will not become either full or empty, thus avoiding nonlinear operation. For negligible friction torque T_s, a simplified relationship is obtained:

$$2^n > 2.14 \frac{f_m}{K} \tag{6-29}$$

The DAC gain K_c depends on its maximum output voltage U_m and the number of stages

$$K_c = \frac{U_m}{2^{n-1}} \tag{6-30}$$

In order to enable bidirectional rotation, the DAC output voltage can vary between $-U_m$ to $+U_m$.

The final step in the design is to determine the drive-unit gain from Eq. (6-10):

$$K_1 = \frac{K}{K_c K_g K_e} \tag{6-31}$$

Note that all the other gains in Eq. (6-31) have been calculated already. Also note that in order to guarantee a full-range operation, the maximum effect input voltage to the servodrive unit should be equal to the maximum operating output voltage of the DAC.

Design example This design procedure is illustrated for the digital loop of an NC system for a high-power lathe [11].

The feed drives selected for the lathe were dc servomotors rated at 120 lb·in (13.6 N·m) nominal torque. Technical specifications of the motor are given in Table 6-1. The lathe is equipped with 10-mm pitch leadscrews and a resolution of BLU = 0.01 mm is required. The maximum required feedrate is 1200 mm/min (~27 in/min), which corresponds to 2000 pps according to Eq. (6-23).

Table 6-1 Technical specification of servomotor

Nominal torque	T	120 in · lb
Nominal speed	ω	720 rpm
Torque constant	K_t	10.27 in · lb/A
Motor constant	K_m	0.862 rad/(s · V)
Mech. time constant	τ_m	11.96 ms
Armature resistance	R	0.75 Ω
Moment of inertia	J	0.19 in · lb · s^2

From Eq. (6-25), it is concluded that an encoder of 1000 pulses per revolution is required. This frequency was achieved with an encoder of only 250 square-waves per revolution by using its two 90° phase-shifted outputs. The two outputs are required for the direction-sensing circuit, and by using the falling and rising edges of both output waves as pulse sources, the encoder fundamental frequency is multiplied by a factor of 4.

For the nominal recommended motor speed of 720 rpm, the required gear ratio from Eq. (6-24) is

$$K_g = \frac{2000 \times 60}{720 \times 1000} = \frac{1}{6}$$

For designing the optimal gain according to Eq. (6-26), the open-loop time constant τ has to be found. The drive unit contains an internal loop, consisting of the amplifier (a PWM type), the motor, and a tachometer as a second feedback device. A typical experimental response of this internal loop is shown in Fig. 6-13.

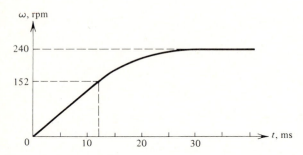

Figure 6-13 Experimental velocity response of the machine drive unit.

For simplicity, this time response is treated like that of a first-order system. The characteristic time constant at 63 percent (152 rpm) of the steady-state output is $\tau = 12$ ms which, from Eq. (6-26), gives $K = 42$ s^{-1}.

At normal operating speeds, the magnitude of the load torque is about 10 percent of the rated torque, i.e., 12 in·lb. Calculation of K_2 according to Eq. (6-3)(assuming $\alpha = 1$) with the data given in Table 6-1 gives $K_2 = 0.063$ rad/(s·in·lb) or 0.4 r/(s·in·lb). The maximum position error in the counter at steady state as given by Eq. (6-27) is calculated as follows:

$$E_{\text{max}} = \left[\frac{2000}{42} + \frac{1000 \times \frac{1}{6} \times 0.4 \times 12}{42} \right] 1.07 = 72$$

$E_{\text{max}} = 72$ dictates an 8-bit counter and an 8-bit DAC according to Eq. (6-28). Actually, such a DAC has a larger capacity than required and, therefore, can accommodate unexpected overloads. A DAC with a maximum output voltage of ± 10 V was chosen yielding a gain of $K_c = 10/128$, according to Eq. (6-30).

The designed system has been constructed and tested on a NC lathe. The experimental results, presented in Fig. 6-14, clearly demonstrate that a practical optimal gain almost coincides with the calculated one of $K = 42$ s^{-1}. Lower gains (e.g., $K = 21$) yield sluggish response, while high gains (e.g., $K = 72$) cause oscillations.

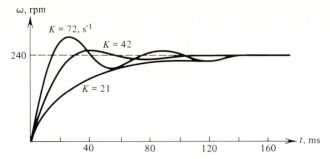

Figure 6-14 Experimental closed-loop velocity response with the open-loop gain as a parameter.

Finally, the output of the DAC, converted to counter pulses, was recorded at low speed. A typical plot taken at 30 mm/min (1.2 in/min) without cutting load is presented in Fig. 6-15. This plot is quite similar to the theoretical one given in Fig. 6-9. However, due to nonconstant friction torques, more variations appear around the average value of the error.

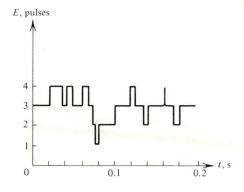

Figure 6-15 Experimental position error at low velocity.

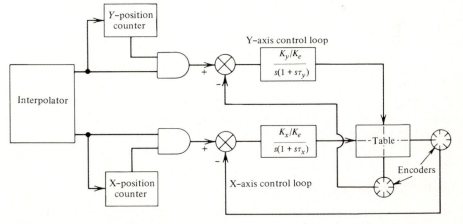

Figure 6-16 Two-axis contouring system.

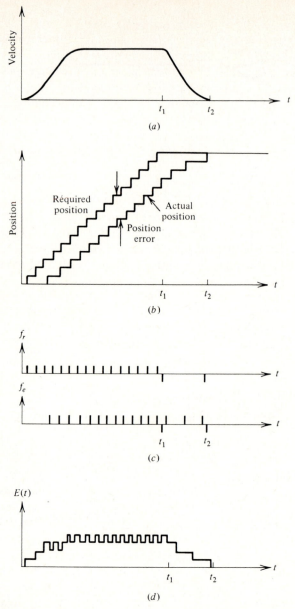

Figure 6-17 Response of a closed-loop system to a constant velocity command. (*a*) Axial velocity; (*b*) the required and actual position of the axis; (*c*) the reference and encoder frequencies; (*d*) the state of the up-down counter in the loop.

6-3.4 Position Control

The input signals to the digital control loops in NC contouring systems are reference pulses generated by the interpolator. Since each pulse causes a motion of one BLU of axis travel, the number of reference pulses can be counted to determine the required positions of each axis in BLUs. The required positions for each axis of motion are given by *position counters,* which are contained in the NC controller, as shown in Fig. 6-16. Each axis of motion is equipped with a position counter, the contents of which represent the *instantaneous required* position of the axis. The *actual* position, however, differs from the one indicated by the position counter. The *position error E* , as illustrated in Fig. 6-17 for a linear motion, is the difference between the required and the actual positions. This position error, which is contained in the axial up-down counter, is accumulated from the start of the motion as shown in Fig. 6-17*d*. For a constant reference frequency f_r, the position error reaches the steady state when the axis runs at the corresponding constant velocity, as shown in Fig. 6-17.

When the position counter reaches the final required position at time t_1, the flow of the reference pulses is blocked by an AND gate as shown in Fig. 6-16. However, since the axis is still in motion, the feedback pulses are sent and reduce the contents E of the up-down counter to zero at time t_2, which indicates that the actual position is equal to the required one and the tool is at the end of the segment. As a consequence, both the position counter and the up-down counter control the position of the axis. By using this control method, the error E is gradually reduced to zero, which, in turn, causes a gradual decrease in the motor velocity and the axis approaches the target point very smoothly under automatic deceleration. It is worthwhile to note that an additional simultaneous deceleration of two or more axes can be achieved by controlling the output pulse frequency of DDA 3 of the interpolator (see Fig. 5-12) as was explained in Sec. 5-2.3.

6-3.5 Operation of a Two-Axis System

From a control point of view, the significant common requirement of all NC and CNC systems is to generate coordinated movement of the separately driven axes of motion in order to trace a predetermined path of the tool relative to the workpiece. A block diagram of a typical NC contouring system with two axes of motion is shown in Fig. 6-16. The coordinated movement is generated by the interpolator which transmits (in NC systems) the reference pulses and by the closed-loop circuit with which each axis is equipped.

The most significant factor in the performance of contouring systems is the accuracy of the overall system. A measure of the accuracy is the *contour error,* which is defined as the distance difference between the required and actual path. Three cases will be considered: (1) constant velocity along a linear path, (2) constant angular velocity around a circular arc, and (3) cutting a corner.

Contour error in linear motion. The actual axial positions in linear motion are given according to Eq. (6-20):

$$X(t) = f_x t - E_x \qquad (6\text{-}32a)$$

$$Y(t) = f_y t - E_y \qquad (6\text{-}32b)$$

where f_x and f_y are either the reference frequencies or the required axial velocities. For a negligible load torque the axial position errors E_x and E_y at steady state are given by

$$E_x = \frac{f_x}{K_x} \qquad E_y = \frac{f_y}{K_y} \tag{6-33}$$

where K_x and K_y are the axial open-loop gains. By eliminating the time t in Eqs. (6-32), the actual path is obtained:

$$Y = \frac{f_y}{f_x} X + \frac{f_y E_x - f_x E_y}{f_x} \tag{6-34}$$

The required path, however, is given by

$$Y = \frac{f_y}{f_x} X \tag{6-35}$$

Comparing Eq. (6-34) with Eq. (6-35) shows that the second term in Eq. (6-34) represents a path error. The shortest distance between the required and the actual path is the contour error ϵ, which is given at steady state by [10, 14]

$$\epsilon = \frac{f_y E_x - f_x E_y}{f} \tag{6-36}$$

where $f = \sqrt{f_x^2 + f_y^2}$ is the required velocity along the path. Substituting Eqs. (6-33) into Eq. (6-36) yields the contour error:

$$\epsilon = \frac{f_x f_y}{fK} \frac{\Delta K}{K} \tag{6-37}$$

where K is the average open-loop gain defined by $K = \sqrt{K_x K_y}$ and $\Delta K = K_y - K_x$. The factor $(\Delta K / K)$ represents the amount of mismatching in the system which creates the contour error. For high contouring accuracy it is important to have both well-matched gains and high average gain, because these two conditions together tend to reduce the contour error.

Example 6-5 A straight-cut milling of an aluminum workpiece is performed at 45° on the XY plane at 450 mm/min. The system resolution is BLU = 0.01 mm. The open-loop gains are adjusted to 15 s^{-1} ± 2 percent. Calculate the maximum contour error.

SOLUTION The velocity along the path is given by

$$f = \frac{450}{60} = 7.5 \text{ mm/s}$$

and $f_x = f_y = f/\sqrt{2}$. According to Eq. (6-37), with a maximum deviation of 4 percent, which yields $\Delta K / K = 0.04$,

$$\epsilon = \frac{7.5 \times 0.04}{2 \times 15} = 0.01 \text{ mm}$$

This shows that the contour error is equal to the system resolution. Such a contour error is the maximum permitted one in many applications.

When the two axes are perfectly matched, i.e., $K_x = K_y$ and $\tau_x = \tau_y$, there is zero contour error, although there are individual axis errors with respect to time. In most practical systems, however, due to a variety of causes, there will be a certain amount of mismatching between the axes' characteristics, resulting in a contour error.

Contour error in circular motion. For a desired circular contour with radius R and perfectly matched system dynamics, the actual steady-state contour generated will also be a perfect circle with radius larger or smaller than the desired radius, depending upon the amount of damping and the angular velocity. The steady-state radial error for a constant angular velocity ω and a perfectly matched system is given by [14]:

$$\frac{e_r}{R} = 1 - \frac{1}{\sqrt{1 + (2\zeta\omega/\omega_n)^2 - 2(\omega/\omega_n)^2 + (\omega/\omega_n)^4}} \qquad (6\text{-}38)$$

where ζ and ω_n are the system damping factor and natural frequency, respectively. To obtain small contour errors the condition $\omega \ll \omega_n$ must be satisfied.

It can be seen that for perfectly matched axes, the amount or radial error is very small, especially when the damping factor of the system is about 0.7. In most contouring systems, there will be a certain amount of mismatch between the axes' dynamics which will affect the error obtained. With a mismatched system the steady-state contour error becomes roughly elliptical in shape and the errors are considerably increased [14].

Contour error in cutting a corner. The error obtained when cutting a corner is also a measure of the system accuracy. Consider the case in Fig. 6-18 in which the tool should cut a corner. The position of the tool along the X axis with respect to time is given in Fig. 6-17. At time t_1 the X position counter indicates the end of the segment, and a motion along the Y axis starts. However, at time t_1 the X axis is still moving since it is in a distance of E_{ss} (BLUs) from the end of the segment, and it will come to a full stop only at time t_2. Therefore during the time period $t_2 - t_1$ both axes are moving, resulting in the shape given in Fig. 6-18.

The shape in Fig. 6-18 has been calculated for a system with $\tau = 20$ ms, open-loop gain $K = 35$ s^{-1} (resulting in $\zeta = 0.6$), and a velocity of 14 mm/s. In this system both axes are moving during a 130-ms interval and result in a maximum overcut and undercut which are almost equal. Usually the undercut is less significant than the overcut, and the actual cutting with a non-zero radius of the tool even reduces the size of the undercut. Therefore a design for $\zeta = 0.7$ which yields a negligible overcut is regarded as optimal.

Figure 6-19 compares a corner cutting operation for the previous system with different values of open-loop gain. If the gain is too low (e.g., $K = 17$ s^{-1}), the position error in the X axis is large and the system is sluggish, which results in a significant size of undercut. In contrast, if the gain is too high ($K = 70$), the cutting is performed with overshoots, causing an overcut and unsmooth surface. The best performance is achieved in this system with $K = 25$ s^{-1}, which provides a relatively fast response, smooth cutting, and a negligible overcut.

Finally it should be emphasized that lower velocities (i.e., machining feedrates) reduce the three types of contour errors: in linear, circular, and corner cutting. In order to have a highly accurate contouring system with high velocities, the system damping should be adjusted around $\zeta = 0.7$ and the system gains must be well matched.

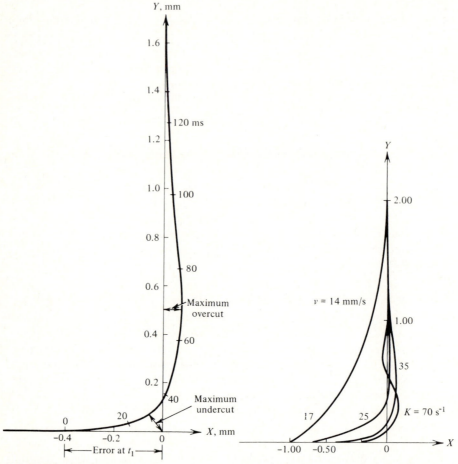

Figure 6-18 Cutting a corner with a manufacturing system ($\zeta = 0.6$).

Figure 6-19 Cutting a corner with various gains: $K = 17$ s^{-1}($\zeta = 0.86$); 25 (0.71); 35 (0.60); 70 (0.42).

BIBLIOGRAPHY

1. Beckett, J. T., and H. W. Mergler: Analysis of an Incremental Digital Positioning Servosystem with Digital Rate Feedback, *Trans. ASME J. Dyn. Sys. Meas. Contr.*, vol. 87, March, 1965.
2. Buckerfield, S. T.: Continuous Numerical Control of Machine Tools, *Control*, vol. 3, pp. 90–98, June/July, 1960.

3. Evans, J. T., and L. V. Kelling: Inside the Mark Century Numerical Control, *Contr. Eng.*, vol. 10, pp. 112–118, May, 1963.
4. Ertell, G.: "Numerical Control," Wiley-Interscience, New York, 1969.
5. Haringx, J. A.: A Numerically Controlled Contour Milling Machine, *Philips Tech. Rev.*, vol. 24, pp. 299–331, September, 1963.
6. General Electric: "Pulse Width Modulated Servo Drive," GEK-36203, March, 1973.
7. Koren, Y., A. Shani, and J. Ben-Uri: Overshoot Correction in Digital Control System, *Control*, vol. 13, no. 129, pp. 204–205, March, 1969.
8. ———, A Shani, and J. Ben-Uri: Numerical Control of a Lathe, *IEEE Trans. Ind. Gen. App.*, vol. 6, no. 2, pp. 175–179, March, 1970.
9. ———: Resolver in Digital Control Loop, *IEEE Trans. Ind. Elec. Contr. Inst.* (feature presentation), vol. IECI-24, no. 2, pp. 145–149, May, 1977.
10 ———: Cross-coupled Computer Control for Manufacturing Systems, *Trans. ASME, J. Meas. Dyn. Contr.*, vol. 102, no. 4, pp. 265–272, December, 1980.
11. ———: Design of a Digital Loop for NC, *IEEE Trans. Ind. Elec. Contr. Inst.*, vol. IECI-25, no. 3, pp. 212–217, August, 1978.
12. ———, and J. Ben-Uri: "Numerical Control of Machine Tools," Khanna Publishers, Delhi, 1978.
13. Olesten, N. O.: "Numerical Control," Wiley-Interscience, New York, 1970.
14. Poo, A. N., and J. G. Bollinger: Dynamic Errors in Type 1 Contouring Systems, *IEEE Trans. Ind. App.*, vol. IA-8, no. 4, pp. 477–484, July/August, 1972.
15. Taft, C. K., F. N. Lutz, and M. Mazoh: Dynamic Accuracy in Numerical Control Systems, Part I, *Tool Manufac. Eng.*, pp. 18-20, May, 1967; Part II, *ibid*, pp. 80–83, June, 1967.
16. Vignon, H. E.: Effect of Servo System Characteristics on the Accuracy of Contouring Around a Corner, *Proc., AIEE,* paper 62-250, 1962.

PROBLEMS

6-1 A drilling machine in which the BLU = 0.02 mm is equipped with stepping motors as the axial actuators (see Fig. 6-2). The required axial feedrate is 2.4 m/min, and a deceleration circuit becomes active 0.4 mm before the target point. This circuit consists of one trigger flip-flop which acts as a frequency divider by 2. Calculate

    (*a*) The input frequency to the stepping motor before the deceleration.

    (*b*) The deceleration time.

6-2 A PTP NC device has a rapid traverse of 3600 mm/min and two-stage deceleration circuit (see Fig. 6-5). The fine feedrate of 360 mm/min starts 1.0 mm before the target position, and the creep feedrate of 36 mm/min starts 0.1 mm before the target position. Calculate the total deceleration time and draw a feedrate-time diagram on log-log paper.

6-3 The punched tape of a contouring NC system contains the following manual program:

    g00 x+1000 y−1000
    g01 x+6000 y−5000
    g01 x+9000 y+8000
    g01 x+1000 y−1000
    g00 x0   y0

The distances are given in absolute dimensions, and the tool is initially at (0, 0).

    (*a*) Make a list of the contents of the command and position registers (Fig. 6-6) at the beginning of each block.

    (*b*) Draw the part; include coordinates.

6-4 A stepping motor of 200 steps per revolution was directly mounted on a leadscrew of 1.5-mm pitch. These values yield a linear motion of 0.0075 millimeters per step. The steps are counted in a down-counter as shown in Fig. 6-20. It is desired that each bit of the counter will represent 0.01 mm. Propose a suitable logic device to compensate for the different numbers and add it in the appropriate place in the diagram.

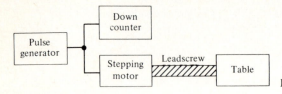

Figure 6-20 Block diagram of Prob. 6-4.

6-5 In a PTP system each axis decelerates to one-half of its normal speed 50 BLUs before its target position. Draw the path of the tool if an incremental motion of $X = 250$, $Y = 200$ BLUs is performed. Note that both axes are moving at the same speed until one of them reaches the deceleration stage.

6-6 Design a single axis open-loop control for a NC drilling machine. The control contains a pulse generator which transmits 2000 pps, a decelerator circuit, and a position down-counter. The counter is loaded to the required distance in BLUs and is decremented during the motion. The stepping motor should be decelerated from a velocity of 2000 steps per second to 250 steps per second about 40 ms before stopping. The motor stops when the counter reaches the zero position. Draw a block diagram of the system.

6-7 The dominant time constant in the digital control loop shown in Fig. 6-10 is 20 ms; the open-loop gain is 35 s^{-1}; the leadscrew pitch is 5 mm; and the encoder gain is 500 pulses per revolution. The input pulses to the loop are transmitted at a constant frequency of 1500 pps. If the load torque is negligible, calculate the following:

(a) The linear velocity and the position of the table at steady state
(b) The damping factor and the natural frequency of the loop
(c) The steady-state position error
(d) The number of binary stages in the counter

6-8 Draw the velocity and the position of the table in Prob. 6-7 during the first 200 ms.

6-9 Calculate the steady-state path error generated by the operation of two perpendicular axes acting in a closed loop at the same velocity. The velocity along the path is 4 mm/s (BLU = 0.01 mm); the open-loop gain is 25 s^{-1} and the difference between the loop's gain is 2 percent. What is the maximum permitted velocity if the error should be smaller than 1 BLU?

6-10 The output velocity of the control loop shown in Fig. 6-8 can be approximated by:

$$v = 10t \qquad 0 < t < 0.1 \text{ s}$$
$$v = 1 \qquad 0.1 < t$$

where v is given in millimeters per second. If the input frequency to the loop is 100 pps, calculate:

(a) The number of pulses the encoder provides for each millimeter of linear motion
(b) The equations for the required and actual position
(c) The position error at steady state (in millimeters)
(d) The average number in the counter at steady state
(e) The open-loop gain
(f) Draw diagrams of the following:
   The input and encoder frequencies
   Required and actual position
   The contents of the up-down counter

SEVEN

COMPUTERIZED NUMERICAL CONTROL

This chapter deals with computerized controllers of manufacturing systems. The first two sections describe the principles of operation of these controllers and their advantages compared with conventional hard-wired NC. The third section offers a general introduction to digital computers and explains terms which are subsequently used in the text. Readers who are familiar with computers might skip this section.

Two types of CNC systems, referred to as reference-pulse and sampled-data, are discussed in Secs. 7-4 and 7-5. In the first system reference pulses are generated by the computer and supplied to an external digital control loop which was discussed in Chap. 6. With the sampled-data technique the control loop is closed through the computer itself, which transmits the position error at fixed time intervals. Each of these CNC types requires a distinct interpolation routine in the control program to generate the corresponding reference signals: pulses or binary words. Interpolators were discussed in Chap. 5; Sec. 5-4 deals with a software interpolator which produces reference pulses and Sec. 5-5 is concerned with interpolators suited for sampled-data systems.

The last section evaluates the role of the microprocessor in CNC equipment and introduces modern systems which contain a programming-aid processor in addition to the control program.

7-1 CNC CONCEPTS

An important advance in the philosophy of NC of machine tools, which took place during the early 1970s, was the shift toward the use of computers instead of controller units in NC systems. This produced both computer numerical control (CNC) and direct numerical control (DNC). CNC is a self-contained NC system for a single machine tool including a dedicated minicomputer controlled by stored instructions to perform some or all of the basic NC functions. With DNC, several machine tools are directly controlled by a central computer. Of the two types of computer control, CNC has become much more widely used for manufacturing systems (e.g., machine tools, welders, laser-beam cutters) mainly because of its flexibility and the lower investment required. The preference for CNC over DNC is continuing to become even greater due to the availability and declining costs of minicomputers and microcomputers.

One of the objectives of CNC systems is to replace as much of the conventional NC hardware with software as possible and to simplify the remaining hardware. There are many ways in which functions can be shared between software and hardware in such systems, but all involve some hardware in the controller dedicated to the individual machine. This hardware must contain at least the servoamplifiers, the transducer circuits, and interface components, as shown in Fig. 7-1.

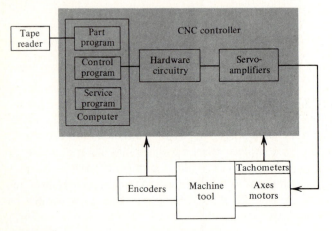

Figure 7-1 Schematic diagram of a CNC system.

The software of a CNC system consists of at least three major programs: a part program, a service program, and a control program. The part program contains a description of the geometry of the part being produced and the cutting conditions such as spindle speed and feedrate. Dimensions in part programs are expressed by integers in units corresponding to the position resolution of each axis of motion. This unit is referred to as the BLU, which might be on the order of 10 μm in a typical machine tool system. The service program is used to check, edit, and correct the part program. The control program accepts the part program as input data and produces signals to drive the axes of motion.

In all types of CNC systems the control program performs interpolation, feedrate control, deceleration and acceleration, and contains position counters which show the incremental distance to the end of the current segment along the required path. The main routine in the control program is the interpolator, which coordinates the motion along the machine axes, which are separately driven, to generate the required machining path. The machining path is usually obtained from a combination of linear and circular segments and accordingly the control program contains a linear and a circular interpolation subroutine.

Most closed-loop CNC systems include both velocity and position control loops, the velocity feedback provided by a tachometer and the position feedback by an encoder or a resolver (see Fig. 7-1). The tachometer provides a voltage proportional to the velocity and the encoder emits voltage pulses, each corresponding to an axis displacement of 1 BLU.

The computer output in CNC systems can be transmitted either as a sequence of reference pulses or as a binary word in a sampled-data system. With the first technique, the computer produces a sequence of reference pulses for each axis of motion, each pulse generating a motion of 1 BLU of axis travel. The number of pulses represents position, and the pulse frequency is proportional to the axis velocity. These pulses can actuate a stepping motor in an open-loop system or can be fed as a reference to a closed-loop system. With the sampled-data technique, the control loop is closed through the computer itself. The control program compares a reference word with the feedback signal to determine the position error. This error signal is fed at fixed time intervals to a DAC, which in turn supplies a voltage proportional to the required axis velocity.

7-2 ADVANTAGES OF CNC

The development of CNC systems has progressed as a result of the rapidly improving capabilities, coupled with falling prices, of small computers, a combination that makes the standard computer an attractive component of NC systems. The trend away from conventional NC to the computer control system means a change from purely hardware-based NC to a software-based system, a change that brings the user a number of advantages:

An increase in flexibility.
A reduction in hardware circuits and simplification of the remaining hardware, as well as the availability of automatic diagnostic programs, brings a subsequent need for fewer maintenance personnel.
A reduction in inaccuracies in manufacturing due to a reduced use of the tape reader.
An improvement in the possibilities for correcting errors in part programs—the *editing* feature.
The possibility of using the computer's peripheral equipment for debugging the edited part program; e.g., a plotter can be utilized for drawing the shape of the part.

Let us elaborate on some of these features. For example, the improved flexibility and reduction in hardware of the CNC as compared with NC systems can be demonstrated by the way in which the systems are modified to suit specific needs. To alter a hardware-based system means rewiring, whereas a modification in a CNC system means reprogramming. In implementing the CNC concept, the control manufacturer can adapt a single design to many machine tools, with customizing implemented in easily changed software; the machine tool builder can buy and service one control for an entire line of machine tools; and the user will end up with more features and options at lower cost.

The reduction in inaccuracies in the manufacturing process stems from the fact that, in many CNC systems, the part program tape is read only once and then stored in the computer memory. When machining a part, the computer presents data in a format similar to that from the tape reader (but without pauses between blocks of information). This means that the part program tape in such systems is used only once per manufacturing series, thus avoiding one of the biggest sources of error in NC systems.

In dealing with errors associated with the part program on the tape, CNC systems offer the user an editing feature. Errors of this type relate to mistakes in cutting conditions, such as speed and feedrate; in compensating for the size of the cutting tool; and in the dimensions of the part itself when they exceed permissible tolerances. Since the data of the part program in a CNC system are stored in the computer's memory, the data can easily be modified if necessary rather than having the tape reprocessed—a costly procedure that is followed in conventional NC systems. This editing feature is of particular importance in the manufacture of small series of products. By adding appropriate software, the corrected part program can be automatically punched on a tape, which can be kept for future use.

The peripheral equipment of a CNC system's computer can help the editing process in that the dimensional errors on the part program tape can be sensed with the aid of a plotter or cathode-ray tube (CRT) display. A computer program, generally referred to as the *editor*, allows communication between the part programmer and the stored part program through the keyboard of a teletypewriter. Prior to the production of a part, its edited part program must be proofed against illegal computer codes, radius errors in circular segments, and dimension differences between the beginning and endpoint of the tool motion. This check is generally provided by a sophisticated editor.

7-3 THE DIGITAL COMPUTER

The computers used with NC are either mini- or microcomputers; both are classified as digital-type computers. All digital computers are capable of executing a limited number of defined operations. Typical operations are to move information from one unit of the computer to another, e.g., to and from the memory and the central processing unit (CPU). Other operations involve arithmetical and logical functions, such as adding two numbers together. The computer is ordered to execute a particular operation by giving it an instruction. The computer can execute only one of these instructions at a time; to perform a required task a sequence of these basic instructions must be executed.

Each basic instruction is represented in the computer as a binary word; i.e., a particular combination of zeros and ones which is decoded by the CPU. While each instruction does not necessarily require the same number of bits to define the required action, the organization of the program memory dictates that all instructions be coded into fixed-length words, some bits of which may be redundant for a particular instruction. A sequential list of instructions to be performed is called a *machine code program* or *object program*. The lowest level language which translates English-like instructions to machine code is called *assembler*.

The physical components of the computer are termed *hardware,* while the programs that control the machine are termed *software*.

7-3.1 Principal Structure

A digital computer includes three major units, as shown in Fig. 7-2:

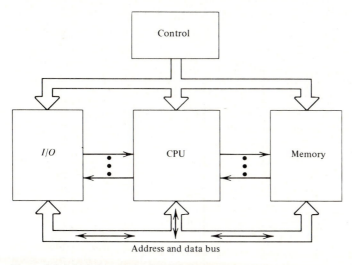

Figure 7-2 Principal structure of a digital computer.

CPU
Memory
Input-output (I/O) section

The flow of data and instructions between these units is controlled by a control unit. The information is transferred in proper timing over a bus (or two buses), which is a common parallel path carrying digital data.

The CPU is the portion of the computer where instructions are decoded and then executed. Data is processed in the CPU where arithmetic or logical types of operations are required in the execution of a program. The memory is the unit which stores the program and data. In addition it can store interim results of the computation. Each instruction is transferred from the memory to the CPU (this only takes around 1 μs) where it is decoded and executed. When an instruction is finished, the next instruction is fetched from the memory and executed. In executing an instruction, data may be involved, and this is made available either by the memory or by an external I/O device.

The I/O portion does exactly what is indicated by its name. It is the place in the system where external digital data is brought in, or data can be output to peripheral devices either serially or in parallel. In short, it is where the interface or connections are made between the computer and the external world.

The number of basic instructions that the computer is capable of executing is directly related to the sophistication of the electronic logic circuitry in the CPU. The heart of the CPU is the arithmetic and logic unit (ALU), which actually performs the desired operations on the data. Typical arithmetic operations are Add, Shift, Compare, and Increment; typical logic operations are AND, OR, Complement, and Clear. Associated with the ALU are a number of storage registers, ranging from 1 to say 16, which act as a scratch pad during arithmetic processing. These registers store data similar to specific memory locations, but they are high-speed transistor-transistor logic (TTL) registers, the data from which can be accessed in tens of nanoseconds rather than the microseconds of core memory.

7-3.2 Computer Memory

Both the instructions and the data are stored and manipulated inside the computer as fixed-length combinations of bits, denoted as *words*. The instructions and data are identified in the memory by individually addressing each word. When a particular address is called, the corresponding word is fetched down into the CPU. Some computers are byte-oriented, where memory is split into individually addressed 8-bit bytes. A 16-bit construction will automatically be fetched as 2 bytes in one operation. In practice all computers use the same basic word length for both data and instructions so that a common memory can be employed; in this way jobs with small programs and large amounts of data, and jobs with large programs and small amounts of data can both be accommodated with a minimum total memory requirement. The accuracy of the computation is directly related to the number of bits in the word.

Typically the program is loaded into one area of memory and data into another. The CPU is initialized by being given the address of the memory location of the first instruction and proceeds from there by the logical progression of the program, referring to data in other specified locations. It must be stressed that there is no physical difference between an instruction and a piece of data; both are similar length binary words. If for any erroneous reason (usually an incorrectly written program) a piece of data is fetched into the CPU when an instruction is expected, then the control section will attempt to decode and execute it with some very puzzling results.

The contents of any memory location, either an instruction or data, must be rapidly available on demand by the CPU. When the memory locations may be referenced in any order at equal time, the memory is said to be a random-access memory (RAM); this can be a magnetic core store or large-scale integration (LSI) semiconductor memory. The magnetic core is slower but maintains the stored information even when the computer is switched off. The semiconductor device is faster but requires power to maintain the information being stored. A non-RAM device such as a magnetic tape and magnetic disk can be used as *backup stores*. A program would be copied from the backup store to memory as a high-speed block transfer and then executed from the memory.

The common feature of all these types of memory is that any information can be read from and written to the device. Another type is the read-only memory (ROM), which is manufactured so that desired data will be read out, and no other data can be written into the device. The data pattern is fixed at manufacturing time. Although a ROM memory also has random-access capability, for some reason the use of the word RAM is commonly applied only to read-write random-access memory. Some of the more common applications for ROMs are: microprocessor program storage, lookup tables, and all types of sequential ROM-driven controllers.

7-3.3 Input and Output

The computer itself is merely an electronic data manipulator. It cannot process data without getting data from the outside (input). Likewise results or control to external devices cannot happen without output. Thus, an important measure of the versatility and utility of a computer is its ability to communicate with external devices or equipment. All the input and output devices supported by the computer are termed *peripherals,* a category which also includes the backup storage devices. Both the nature and number of I/O devices are varied, dependent on the application. One computer can service a number of I/O devices due to the high-speed differences between the essentially elecromechanical peripherals and the electronic processor.

Typical input devices are keyboards on Teletypes, card readers, paper tape readers, and analog-to-digital converters (ADC). Typical output devices are Teletype printers, line printers, card punches, paper tape punches, visual display screens, and DACs.

A word of information, say 16 bits as an example, can be transferred into the computer in one of two modes, either serial or parallel. Serial mode means that the word is transferred 1 bit at a time along a single path in 16 time intervals, the 16 bits being reformed as one word before being used. Parallel mode means that 16 paths are provided and the whole word is transmitted in one time interval. Clearly serial mode is cheaper but slower than parallel mode. One commonly encountered example of serial transmission of data is the Teletype. The character is formed in the Teletype and transmitted along a single line 1 bit at a time, usually a total of 11 bits per 8-bit character, including 3 markers. The string of bits is collated in a shift register at the computer end and only then transferred to the machine in parallel.

One of the simplest configurations for handling I/O is to use a special I/O register. Whenever a data word is ready for output, it is transferred from memory into this I/O register. Or, conversely, when an external peripheral device needs to input a word into the CPU, it merely puts the word in its I/O register and sets its status to indicate that data is ready to be transferred into the computer. ADCs and DACs are usually interfaced with the computer through I/O registers.

Modes of operations. Digital computer systems can be considered to function in one of two modes of operation:

1. *Off-line* The data to be processed is prepared away from the computer, e.g., punched tape or cards, and is brought to the computer for processing at some later time; also termed *batch processing*.
2. *On-line* The data to be processed is communicated directly to the computer system. The computer is dedicated to servicing any requests made by external events as rapidly as possible; results of computations may be returned fast enough to control the condition of the system; also termed *real-time operation*.

The computers which are used at the NC part program preparation stage function in off-line mode, while computers in CNC systems operate in the on-line mode.

The use of computers in on-line mode created a need for a new capability which is not required in off-line applications. In the on-line mode, each external device might send a signal to the computer asking for an immediate service. This signal, which is denoted as an *interrupt,* stops the execution of the current program and starts the execution of a different routine which serves the interrupt. Upon completing the routine the computer continues to execute the original program. By transmitting interrupts at a constant rate, an interrupt service program can be executed at constant time intervals, a feature which is used in CNC systems.

7-4 THE REFERENCE-PULSE TECHNIQUE

In designing a CNC system for a particular application, the first step is to determine whether the reference-pulse technique or sampled-data technique is most appropriate. It will be seen that the choice is constrained in most cases by the velocity and system response requirements.

A block diagram of a single-axis CNC system using the reference-pulse technique is presented in Fig. 7-3. In a practical multiaxis system, each axis of motion is

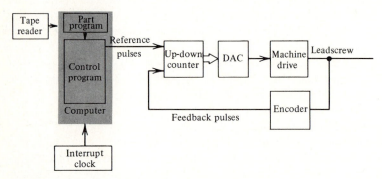

Figure 7-3 Block diagram of a reference-pulse CNC system.

controlled by an identical control loop connected to the computer by two pulsed lines, one for clockwise and the other for counterclockwise motion. The computer contains the control program which accepts the part program as input and produces the reference pulses to the control loop. The machine drive block in Fig. 7-3 includes the axis motor, tachometer, and servoamplifier as shown in Fig. 4-11.

The digital encoder in Fig. 7-3 is the feedback device, and the up-down counter is a comparator element fed by two sequences of pulses: reference pulses from the computer and feedback pulses from the encoder. The pulse number difference between the two inputs is the position error, which is converted to a voltage by a DAC to drive the axis motor. The motor rotates in the direction which reduces the error. For example, if a constant reference frequency is supplied by the computer, the encoder frequency in the steady state is identical to the reference frequency, except for a finite pulse and phase difference which is necessary to generate the corrective error voltage to rotate the motor. An alternative method uses a resolver as the feedback device and a phase comparator, or a discriminator, to compare the reference and the feedback signals. In either case, the input to the control loop is a sequence of reference pulses, each pulse generating a motion of 1 BLU, so that the axis velocity is proportional to the pulse frequency and its position proportional to the number of transmitted pulses.

The interpolator in reference-pulse systems is based upon an iterative technique controlled by an external interrupt clock. At each interrupt a single iteration of the interpolator routine is executed, which in turn can provide an output pulse to one or more axes. Such a typical interpolator is the software DDA which was introduced in Chap. 5. The maximum output pulse rate is obtained when each iteration of the interpolator produces an output pulse, so that the interrupt frequency f_m for a maximum velocity is

$$f_m = \frac{V_m}{\text{BLU}} \qquad (7\text{-}1)$$

where V_m is the maximum axis velocity in BLU units per second.

The maximum allowable interrupt rate depends on the execution time of a single iteration of the interpolator routine. With present-day technology, the execution time t_e is approximately 100 μs, which in turn yields a maximum interrupt frequency of

$$f_m = \frac{1}{t_e} = 10^4 \text{ pps} \qquad (7\text{-}2)$$

This is also the maximum frequency of reference pulses in interpolated motions. It should be noted that when applying rapid traverse motions which do not require coordination between axes of motion, the major portion of the interpolator routine is bypassed and, therefore, the frequency of the reference pulses can be two or three times faster than that given by Eq. (7-2). Combining Eqs. (7-2) and (7-1) gives the relationship between the system BLU and the maximum axis velocity which can be achieved by the reference-pulse technique. If the maximum velocity requirement exceeds this limit, it will be necessary to use the sampled-data technique.

In principle, the reference-pulse technique should be treated mathematically as a discrete-time system with nonuniform sampling, in which the continuous motion of the axis is monitored from pulses emitted K_e times per revolution of the encoder. In practice, however, since K_e is relatively large (e.g., 1000 pulses per revolution), the reference-pulse system can be analyzed as a continuous system, to which the Laplace transform technique can be applied for analyzing the design requirements. The transfer function of the machine drive unit can be modeled as a first-order system (see Eq. 6-2):

$$G_1(s) = \frac{K_1}{1 + s\tau} \tag{7-3}$$

where τ is the dominant time constant of the machine drive, and K_1 is the machine drive constant given in velocity units per volt. The up-down counter in Fig. 7-3 converts the frequencies to a pulse number and phase difference, both being the integral of frequency. Therefore, the counter functions as an integrator in the loop, and its transfer function is given by

$$G_2(s) = \frac{1}{s} \tag{7-4}$$

The open-loop gain of the system is

$$K = K_c K_1 K_e \tag{7-5}$$

where K_c is the DAC gain (volts per pulse) and K_e is the encoder gain (pulses per revolution). Combining Eqs. (7-3) through (7-5) gives the open-loop transfer function

$$G_0(s) = \frac{K}{s(1 + s\tau)} \tag{7-6}$$

and consequently the closed-loop transfer function is

$$G(s) = \frac{K}{\tau s^2 + s + K} \tag{7-7}$$

It is seen from Eq. (7-7) that the closed loop behaves as a second-order system with the characteristic equation:

$$s^2 + 2\zeta\omega_n s + \omega_n^2 = 0 \tag{7-8}$$

where the damping factor is

$$\zeta = \frac{1}{2\sqrt{K\tau}} \tag{7-9}$$

and the natural frequency is

$$\omega_n = \sqrt{\frac{K}{\tau}} \tag{7-10}$$

Since the open-loop system contains an integrator, the steady-state velocity error of the closed-loop system to a step input is zero. A step input in the reference-pulse system means a constant reference frequency supplied by the computer. Typical velocity responses for various open-loop gains taken on an actual CNC drive with $\tau = 12$ ms are given in Fig. 6-14. Notice that although the steady-state error is always eliminated, the maximum overshoot is increased with the open-loop gain. By decreasing the gain, the maximum overshoot can be reduced at the expense of having a longer settling time. Selecting the open-loop gain as

$$K = \frac{1}{2\tau} \qquad (7\text{-}11)$$

was shown in Chap. 6 to give satisfactory performance with contouring systems. This results in a damping factor $\zeta = 0.707$ which is regarded as optimal for many control systems.

The positional control is performed by the control loop and software position counters, which are contained in the control program. The counters are loaded with the required incremental distance at the beginning of a segment. Each axis of motion is provided with a software position counter. Each time a reference pulse is sent by the computer, the contents of the appropriate counter are decremented by one unit. The up-down counters in the control circuits are never in saturation, thus accomplishing the positional control.

A typical CNC control program of the reference-pulse technique contains five routines:

Traverse In a traverse motion the axes move in the highest allowable feedrate. An interrupt pulse activates the traverse routine, and then henceforth it starts to run in a software loop supplying reference pulses, the frequency of which depends on the cycle time of this loop. A decrementation of the appropriate position counter by one unit is carried out for each reference pulse. An automatic deceleration is accomplished when the contents of a counter are close to zero. When all counters are reaching zero, the control program continues.

Feedrate The feedrate routine generates interpolation commands in a rate dependent on the feed word (f) in the corresponding block of the part program. The maximum rate of interpolation commands is equal to the frequency of the interrupt pulses. If the block contains a deceleration command, the feedrate decreases exponentially to a fixed percentage of the programmed feedrate in that block.

Interpolator Three types of interpolation are usually available: linear and two circular interpolations. The principle of each interpolator is a software simulation of DDA integrators (see Chap. 5). For every interpolation command which is produced by the feedrate routine, a single cycle of the DDAs is simulated.

Output If as the result of a DDA cycle an overlow pulse is generated either in one or more axes, this routine sends the reference pulses.

Position For every reference pulse the position counter of the appropriate axis is decremented by one unit. A zero position check of all counters is performed. When they are at zero, which means that the machining of the current segment has terminated, the control program loads a new block from the part program.

7-5 SAMPLED-DATA TECHNIQUE

The sampled-data CNC system used the computer as part of the control loop by replacing the up-down hardware counter in Fig. 7-3 with a software comparator. A block diagram of the sampled-data system for a single axis is presented in Fig. 7-4.

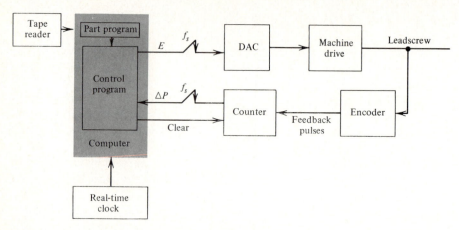

Figure 7-4 Block diagram of a sampled-data CNC system.

The computer samples the feedback signal at a constant frequency f_s and compares it with a reference produced by the interpolator. The resulting error is the computer output which is transmitted at a uniform rate through a DAC to the drive motor. Although an incremental encoder is used in the system illustrated in Fig. 7-4, a resolver or an inductosyn could be substituted. The required interface circuitry will depend upon the hardware chosen. With an incremental encoder the interfacing is simplest, consisting of a small counter which is incremented by the pulses received from the encoder. The computer samples the contents of the counter ΔP at fixed time intervals T and immediately clears it. Since each pulse from the encoder is equivalent to 1 BLU, the number transferred from the counter to the computer is equal to the incremental axis displacement in BLUs during the last period T. *The interval T is adjusted by the real-time clock.*

7-5.1 Design Principles

The control program is based upon an iterative technique which is executed at fixed sampling rate f_s, where $f_s = 1/T$. Since the sampled-data technique applies a complicated interpolation routine, the execution time of a single iteration in a sampled-data system is much longer than that obtained by the reference-pulse technique. At each iteration, the contents of the counter ΔP are accumulated in the control program:

$$P(n) = P(n-1) + \Delta P(n) \tag{7-12}$$

where P represents the actual position of the axis of motion in BLUs and is subtracted from the reference R to generate the position error E:

$$E(n) = R(n) - P(n) \tag{7-13}$$

The error is converted by the DAC and fed to the machine drive. R is the position reference and is related to the required velocity V by

$$R(n) = R(n - 1) + K_eTV(n) \tag{7-14}$$

Equations (7-12) to (7-14) yield an alternative method to calculate the error:

$$E(n) = E(n - 1) + K_eTV(n) - \Delta P(n) \tag{7-15}$$

The value of $TV(n)$ is determined on-line by the interpolator routine. Equation (7-15) presents an iterative process for carrying out numerical integration within the control loop, and consequently the steady-state velocity error of the closed-loop to a step input is zero.

The velocity response $W(n)$ of the axis to a constant velocity reference of V units is given by [7]:

$$W(n) = V[1 - e^{-n\alpha T}(\cos n\omega_1 T + M \sin n\omega_1 T)] \tag{7-16}$$

where

$$\alpha T = - \ln(a) \tag{7-17}$$

and

$$\omega_1 T = \arccos \frac{1 + X - KT + K\tau(1 - X)}{2a} \tag{7-18}$$

The parameters X, a, and M, are defined by

$$X = e^{-T/\tau}$$

$$a^2 = X + K [\tau(1 - X) - TX]$$

$$M = \frac{(1 - X)(1 + K\tau) - KT}{2a \sin \omega_1 T}$$

where τ is the time constant of the machine drive unit and K is defined in Eq. (7-5). The damping factor in the sampled-data system is given by

$$\zeta = \frac{\alpha T}{\sqrt{(\alpha T)^2 + (\omega_1 T)^2}} \tag{7-19}$$

Equation (7-16) shows that at the steady state $W(n) = V$, and the rate at which the steady state value is approached is directly dependent on how quickly the term $e^{-n\alpha T}$ goes to zero which, in turn, depends on K and T. For small K and large T the velocity response approaches its steady state value slowly (a sluggish system). On the other hand, if K is too big the system will oscillate before approaching the steady state value. This behavior can be seen in Fig. 7-5 which shows the velocity response and the position error E of a sampled-data system with a big open-loop gain of $K = 65$ s^{-1}. This result was obtained on a hybrid computer with the machine drive unit simulated according to Eq. (7-3) and with time constant $\tau = 10$ ms and sampling frequency $f_s = 50$ Hz.

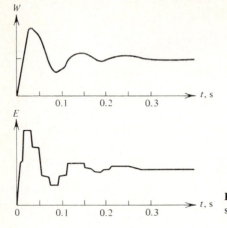

Figure 7-5 Velocity response and position error of a sampled-data system with too high open-loop gain.

One of the prime considerations in using a sampled-data system is choosing the sampling period T and the open-loop gain K. There is always a trade-off between T and K: a longer sampling period requires lower gain in order to maintain a given level of performance. One performance criterion proposed for sampled-data systems with step inputs is minimization of the integral of the absolute value of the error (IAE) [7]:

$$IAE = \int_0^\infty | E(t) | \, dt \tag{7-20}$$

This criterion takes into account the two contrary design requirements of servosystems, namely, small overshoots on the one hand and minimum steady state position errors on the other.

The optimal locus which minimizes the IAE in Eq. (7-20) is included in Fig. 7-6 [7]. Selection of the operating point on this optimal locus depends upon the time constant τ. It is known from the sampling theorem that the sampling rate should be at least twice as fast as the highest frequency ω_0 contained in the sampled signal:

$$\frac{1}{T} = f_s \geq 2\left(\frac{\omega_0}{2\pi}\right) = \frac{\omega_0}{\pi} \tag{7-21}$$

The machine drive of the CNC sampled-data system behaves like a low-pass filter with a critical frequency of $\omega_c = 1/\tau$, as can be seen from Eq. (7-3). Substituting $\omega_0 > \omega_c$ in Eq. (7-21) yields the necessary condition for selecting the sampling period:

$$\frac{T}{\tau} < \pi \tag{7-22}$$

The effect of T is illustrated in Fig. 7-7, which shows the step response of a simulated CNC system with $\tau = 20$ ms with two sets of K and T which lie on the optimal IAE locus of Fig. 7-6. A step torque is applied at $t = 0.1$ s, which might simulate the start of a machining process. With $T/\tau = 0.5$, the response is smooth and fast, but with

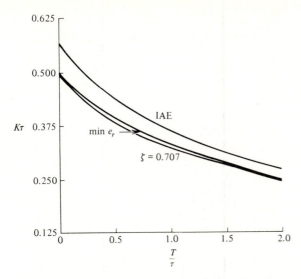

Figure 7-6 Optimal loci of sampled-data systems according to IAE, minimum radial error (min e_r), and $\zeta = 0.707$.

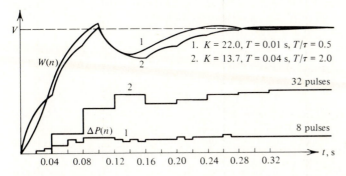

Figure 7-7 Responses and contour readings of a simulated sampled-data system.

$T/\tau = 2.0$, the performance is poorer because the system needs a longer time to recover from the torque disturbance and the motion is rougher. For systems we have studied, satisfactory performance has been obtained with

$$\frac{T}{\tau} \simeq \frac{\pi}{4} \qquad (7\text{-}23)$$

which satisfies the condition in Eq. (7-22) and also assures that the signal's energy loss is less than 15 percent while sampling [3].

7-5.2 Optimization for Circular Motion

Up to this point, the system design has been considered for the case of linear motion. However, CNC manufacturing systems also require nonlinear motion which is generated as a combination of lines and circles.

Circular curves are generated in CNC systems by feeding sinusoidal varying references into the control loops, but the control loops do not follow the sinusoidally inputs perfectly with zero error along each axis. Since a plane circular path is generated by the simultaneous motion of two axes, the radial error depends on the axial errors, which in turn become greater with faster angular velocities. It has been shown [7] that the steady-state radial error with a sampled-data CNC system can be written

$$\frac{e_r}{r} = \frac{L}{2}(\omega T)^2 + \frac{N}{2}(\omega T)^4 \tag{7-24}$$

where r is the radius of the circle, ω is the required angular velocity around the arc, and the parameter L is given by

$$L = \frac{K\tau(T/\tau + 2) - 1}{(KT)^2} \tag{7-25}$$

The parameter N also depends on K, T, and τ. Since in practice $\omega T \ll 1$, the term containing N becomes negligible except when $L = 0$.

For a given circular path, the maximum radial error e_{rm} occurs at the maximum path velocity V_m corresponding to the angular velocity

$$\omega_m = \frac{V_m}{r} = (\text{BLU})\frac{f_m}{r} \tag{7-26}$$

Combining this with Eq. (7-24) and neglecting the second term leads to

$$\frac{e_{rm}}{\text{BLU}} = \frac{L}{2}(Tf_m)^2 \frac{\text{BLU}}{r} \tag{7-27}$$

In order to keep the diametral error within 1 BLU, the radial error should not exceed $\frac{1}{2}$ BLU so that

$$\frac{r}{\text{BLU}} > L(Tf_m)^2 \tag{7-28}$$

Therefore the error constraint leads to a limit on the minimum allowable radius, which is expressed here in BLUs as a function of $K\tau$, T/τ, and τf_m for a sampled-data system. With $Tf_m = 100$ (a practical value [8]) and introducing the minimum IAE criterion from Fig. 7-6, the minimum allowable radius can be more simply expressed as a function of a single parameter, e.g., τf_m. This result is included in Fig. 7-8.

The minimum radius in a system applying the reference-pulse technique is obtained by substituting $T = 0$ in Eq. (7-28), which yields.

$$\frac{r}{\text{BLU}} > (2K\tau - 1)\left(\frac{f_m}{K}\right)^2 \tag{7-29}$$

For $T = 0$ the minimum IAE criterion results in the value of $K\tau = 0.567$ as the optimal, and consequently Eq. (7-29) is reduced to

$$\frac{r}{BLU} > 0.4 \, (\tau f_m)^2 \qquad (7\text{-}30)$$

The corresponding minimum radius has been obtained and is also included in Fig. 7-8.

In many practical CNC systems the minimum allowable radius according to the minimum IAE criterion is too large. Consider, for example, a sampled-data system with a resolution BLU = 0.01 mm and a maximum velocity $V_m = 100$ mm/s. The corresponding f_m according to Eq. (7-1) is 10,000 pps. For a typical time constant of 20 ms, the parameter τf_m becomes 200 and the corresponding minimum radius, according to Fig. 7-8, is 219 mm (8.6 in). Therefore, the minimum IAE criterion and the corresponding design procedure are restricted to systems requiring essentially linear motions.

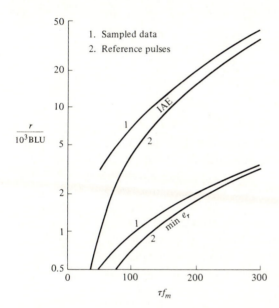

Figure **7-8** Minimum allowable radii in CNC systems for various design criteria.

For CNC systems which require both linear and circular motions, a different design criterion is necessary.

An alternative optimization criterion for design of CNC systems which include a circular interpolator could be minimization of the radial error. For a sampled-data system, the radial error is given by Eq. (7-24). In practice $\omega T \ll 1$ and the first term will usually give the major contribution to the total radial error. This first term can be eliminated, and therefore near-optimal performance can be simply obtained by setting $L = 0$, which from Eq. (7-25) leads to:

$$K = \frac{1}{T + 2\tau} \qquad (7\text{-}31)$$

This result is shown as the curve designated min e_r in Fig. 7-6. It can be seen in Fig. 7-8 that this criterion allows for much smaller radii of curvature than the minimum IAE criterion.

The analysis for a reference-pulse system can be carried out by setting $T = 0$. In Eq. (7-31) this leads to

$$K = \frac{1}{2\tau} \tag{7-32}$$

which is identical to Eq. (7-11) and therefore corresponds to having a damping factor $\zeta = 0.707$. The radial error for a reference-pulse system can be written as [11]

$$\frac{e_r}{r} = 1 - \frac{1}{\sqrt{1 + (2\zeta\omega/\omega_n)^2 - 2(\omega/\omega_n)^2 + (\omega/\omega_n)^4}} \tag{7-33}$$

By combining Eqs. (7-9), (7-10), (7-32), and (7-33), the minimum radial error becomes

$$\frac{e_r}{r} = 1 - \frac{1}{\sqrt{1 + 4(\omega\tau)^4}} \simeq 2(\omega\tau)^4 \tag{7-34}$$

Substituting the maximum value of ω from Eq. (7-26) and limiting the radial error to $\frac{1}{2}$ BLU, the minimum allowable radius is

$$\frac{r}{BLU} = 4(\tau f_m)^{4/3} \tag{7-35}$$

It can be seen in Fig. 7-8 that this minimum radial error criterion allows for a much smaller radius of curvature than the minimum IAE criterion and is slightly smaller than that which can be obtained with the sampled-data technique using the same criterion.

As a general rule of thumb, the damping factor for optimal dynamic performance of many control systems is considered to be $\zeta = 0.707$. The locus corresponding to this damping factor is included in Fig. 7-6 for comparison with the design recommendations according to the present optimization criteria. It can be seen that the minimum radial error criterion gives an open-loop gain with the sampled-data technique that is almost identical to the value for $\zeta = 0.707$, and with the reference-pulse technique $(T = 0)$ an open-loop gain that is identical to that for $\zeta = 0.707$. Therefore, designing the CNC system according to the minimum radial error criterion will also ensure that the damping factor is optimal.

7-5.3 Summary of Design Considerations

The reference-pulse and sampled-data CNC control techniques for manufacturing systems have been described and analyzed. For designing a reference-pulse system, the main parameter is the open-loop gain K, and its optimal value is given in Eq. (7-11) as $K = 1/(2\tau)$. Sampled-data systems require the consideration of two main parameters: the sampling period and the open-loop gain. The sampling period T depends on the time constant τ of the machine drive, and its recommended value according to Eq.

(7-23) is $T \simeq 0.8\tau$. Selection of the open-loop gain is proposed in Eq. (7-31) and dictates a smaller gain compared with the reference-pulse technique.

The reference-pulse technique is much simpler to program, but there is a restriction on the maximum velocity imposed by the control program execution time. By contrast, the maximum velocity in sampled-data systems is not limited by the computer, but the control program is more complex and the open-loop gains are smaller, which results in larger axial position errors and, in turn, bigger contour error in the produced part.

7-6 MICROCOMPUTERS IN CNC

The biggest new force for change in the rapidly growing NC industry was the arrival of the microprocessor. By adapting the technology of the large-scale integration (LSI) to data processing circuitry, a complete 8-bit or 16-bit word microprocessor can be built on a single chip and do the job of dozens of TTL digital circuits with better reliability. A complete microcomputer, built around the microprocessor, can be formed on a single printed-circuit board at very low cost and operate as the main unit of the CNC controller. The increased flexibility, the low cost, and mainly the reliability that comes from using a small number of digital circuits are the dominant characteristics of the microprocessor which will probably encourage the replacement of most minicomputers and all hard-wired controllers by microprocessors. The utilization of the microprocessor in CNC equipment reflects the current trend of using software techniques to minimize or replace hardware, and will probably signify the NC trend of the middle eighties.

7-6.1 The Microprocessor

A microprocessor is the control and CPU of a small computer. Like all computer processors, a microprocessor can handle both arithmetic and logic data under control of a program. To make a computer based on a microprocessor requires the addition of memory (for both the control program and data), time-base logic, and a set of I/O interface circuits to communicate with peripheral devices. Usually the microprocessor works with two types of memories: ROM, in which the control program is stored, and RAM, which is a read-write memory used for the part program and data storage. Microcomputer memory is available in 4K and 8K blocks, where K means 1024 words in the computer world. Obviously the development of the microprocessors must be accompanied by the development of matched memory and logic circuits in order to form a completely self-contained microcomputer. High-speed ROMs and RAMs are examples of specialized circuits; others include I/O buffers, priority interrupt controls, and special clocks to provide several timing phases.

The microprocessors provide an inexpensive computer control to a variety of applications. The cost savings are not limited to component costs, but they extend to other system hardware costs. Connectors can be decreased in number, cabling can be simplified, and cooling requirements are decreased since they need a much smaller power supply. Associated indirect costs also fall, since assembly time is decreased, inventory requirements are reduced, documentation is simpler, and maintenance is easier.

7-6.2 Microprocessors in CNC Systems

A detailed report on the NC industry was issued in September, 1974, by the General Accounting Office (Washington, D. C.).[†] While the report outlines numerous advantages of NC, it also noted that "NC has its disadvantages: It is expensive and complex. Control systems contain thousands of solid-state electronics devices. Such a complexity compounds maintenance problems."

The use of microprocessors in CNC systems provides a solution to the problems of cost and complexity. By employing microprocessors, the number of components in the CNC systems is reduced over that of hard-wired logic systems, which, in turn, results in higher reliability of microprocessor-based systems. Moreover, the microprocessor enables the design of a modular control, whereby each axis of motion of the machine tool requires only one printed circuit board. This scheme minimizes problems of complexity and maintenance.

CNC systems offered today are at a stage where a small machine shop can order one with no fear of running into problems trying to make it work. However, the machine shop must either have a direct or indirect access to a computer center, or use programming services, to fulfill its part programming demands economically. This by no means decreases the effectiveness of the NC equipment. What is obviously wanted is an inexpensive self-contained system, which would be fed by data from the blueprint and be capable of producing the part.

CNC systems which utilize interactive graphics to display 2-D parts and produce them are available for turning. The operator of such a lathe can simultaneously display the profile of the finished part, the actual tool path, and the corresponding part program in alphanumeric codes. A similar system for a laser-beam cutter has been built at the Technion, Israel Institute of Technology. Figure 7-9a shows a computer display of a required part. The design of the part may be modified in the computer and displayed again on the CRT screen. Since the same computer is also used to control the laser-beam cutter, the part can be immediately produced. A picture of the corresponding real part is shown in Fig. 7-9b.

Systems containing a built-in processor which provides automatic programming are also available for lathes. In these systems the operator has to insert the code of the tool, the dimensions of the final part, and the diameter of the workpiece, and the processor calculates the entire path of the tool center. This method minimizes the part program length and programming efforts and can substitute for computer-assisted programming languages such as APT and COMPACT II. Nevertheless, its major drawback is that the machine itself is idle during the programming process.

A CNC system containing a programming processor for turning was developed at the Technion [4] for the lathe shown in Fig. 2-1. To demonstrate the capability of this system, consider the part of which a drawing is shown in Fig. 7-10a. The contour of

---

[†]"American Metal Market," MN Ed., vol. 81, no. 190.

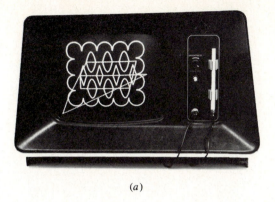

(a)

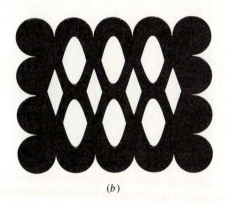

(b)

Figure 7-9 A microcomputer-based interactive system for a laser-beam cutter. (a) Display of the profile of the finished part; (b) the part. *(The system was developed at the Technion, Haifa, Israel.)*

this part consists of four segments. Accordingly, only four programming statements are required to describe the part to the computer. In addition, the computer is fed with the raw workpiece diameter and material, the tool code, and the maximum permissible depth of cut. Based on this information the computer calculates the optimal path of the tool, including all rough cuttings and the finish cut. This path is subsequently displayed on the computer screen, as shown in Fig. 7-10b. Simultaneously the corresponding manual part program can be optionally printed. This program includes 96 manual NC blocks for the part shown in Fig. 7-10a. The computer also calculates the optimal feed and cutting speed based on Taylor's tool-life equation and the tool and workpiece material. Since the programming computer also controls the lathe, the part can be immediately produced. A picture of the finished part is shown in Fig. 7-10c.

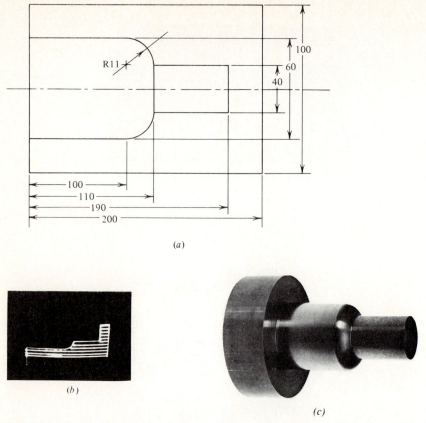

(a)

(b)

(c)

Figure 7-10 A part, which was programmed and produced by a modern CNC system, requires only 4 programming statements instead of 96 NC blocks. *(a)* The drawing of the part; *(b)* the computer display of the tool path; *(c)* the part. *(The system was developed at the Technion, Haifa, Israel.)*

The realization of a full computer-aided design/computer-aided manufacturing (CAD/CAM) system, in which blueprints and punched tapes are eliminated since the data from the design is directly fed to the CNC computer, seems to be closer than ever with the introduction of the microprocessor into the CNC equipment.

BIBLIOGRAPHY

1. Bollinger, J. G.: Computer Control of Machine Tools, *CIRP Ann.*, vol. 21/2, 1972.
2. ———, and J. Mills: The Role of Microprocessors in Future CNC Systems, *CIRP Ann.*, vol. 25/1, pp. 323–328, 1976.
3. Cadzow, J. A.: "Discrete-Time Systems," Prentice-Hall, Inc., Englewood Cliffs, New Jersey, 1973.
4. Green, M.: "Computerized Control to a Lathe and to the Turning Process," M. Sc. Thesis, Technion, July, 1979.
5. Koren, Y.: Design Concepts of a CNC System, *Proc. IEEE Ind. App. Soc., 10th Ann. Meeting,* Atlanta, pp. 275–282, October, 1975.

6. ———: Computer-Based Machine-Tool Control, *IEEE Spectr.* vol. 14, no. 3, pp. 80–84, March, 1977.
7. ———, and J. G. Bollinger: Design Parameters for Sampled-Data Drives for CNC Machine Tools, *IEEE Trans. Ind. App.*, vol. IA-14, no. 3, pp. 255–264, May, 1978.
8. ———: Design of Computer Control for Manufacturing Systems, *Trans. ASME, J. Eng. Ind.*, vol. 101, no. 3, pp. 326–332, August, 1979.
9. ———: Cross-coupled Computer Control for Manufacturing Systems, *Trans. ASME, J. Dyn. Sys. Contr.*, vol. 102, no. 4, December, 1980.
10. Middleditch, A. E.: Design Criteria for Multi-axis Closed Loop Computer Numerical Control Systems, *Trans. ASME, J. Dyn. Sys., Meas. Contr.*, vol. 96, no. 1, pp. 36–40, March, 1974.
11. Poo, A., J. G. Bollinger, and W. Younkin: Dynamic Errors in Type 1 Contouring Systems, *IEEE Trans. Ind. App.*, vol. IA-8, no. 4, pp. 477–484, 1972.
12. Reed M., and H. W. Mergler: A Microprocessor-Based Control System, *IEEE Trans. Ind. Elec. Contr. Inst.*, vol. IECI-24, no. 3, pp. 253–257, August, 1977.
13. Texas Instruments: "TM990—Introduction to Microprocessors, Hardware and Software," Texas Instruments, Dallas, 1979.
14. Willete, E. J.: The Computer's Role in Numerical Control, *Mfg. Eng.*, pp. 36–37, September, 1977.

PROBLEMS

7-1 What is the main difference between CNC and DNC systems?

7-2 What are the advantages of CNC over NC systems?

7-3 Make a list of the different types of computer memory devices.

7-4 The execution time of the interpolator routine in a reference-pulse system is 180 μs, and the resolution is BLU = 0.0001 in. Calculate the maximum axial velocity in inches per minute.

7-5 The required damping factor in a reference-pulse system is $\zeta = 0.707$, and the dominant time constant of the open loop is $\tau = 20$ ms. Select the open-loop gain which guarantees satisfactory performance of the closed loop.

7-6 Repeat Prob. 7-5 for a sampled-data system. The sampling period is $T = 15$ ms.

EIGHT

ADAPTIVE CONTROL SYSTEMS

The adaptive control of metal-cutting processes is a logical extension of the CNC systems. In CNC systems the cutting speed and feedrates are prescribed by the part programmer. The determination of these operating parameters depends on experience and knowledge regarding the workpiece and tool materials, coolant conditions, and other factors. By contrast, the main idea in adaptive control is the improvement of the production rate, or the reduction of machining costs, by calculation and setting of the optimal operating parameters during the machining itself. This calculation is based upon measurements of process variables in real time and is followed by a subsequent on-line adjustment of the operating parameters subject to machining constraints in order to optimize the performance of the overall system.

8-1 INTRODUCTION

In the past decade the number of CNC systems has grown tremendously in almost every field of manufacturing. A common drawback of most of these systems is that their operating parameters, such as speeds or feedrates, are prescribed by a part programmer and consequently depend on his or her experience and knowledge. In order to preserve the tool, even under the most adverse conditions (which in reality will seldom occur), the part programmer prefers to select conservative values for the operating parameters which consequently slow down the system's production.

The availability of a dedicated computer in the control system and the need for higher productivity has greatly accelerated the development of adaptive control (AC) systems for manufacturing. These systems are based on automatic control of the operating parameters with reference to measurements of the machining process vari-

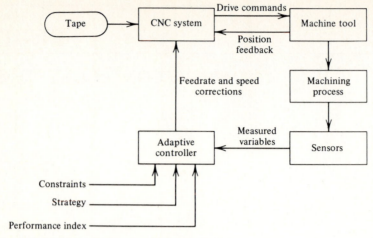

Figure 8-1 Adaptive control system for machine tool.

ables. The adaptive control is basically a feedback system (see Fig. 8-1), in which the operating parameters automatically adapt themselves to the actual conditions of the process.

AC systems for machine tools can be classified into two categories: (1) adaptive control with optimization (ACO), and (2) adaptive control with constraints (ACC). ACO refers to systems in which a given *performance index* (usually an economic function) is extremized subject to process and system *constraints*. With ACC, the machining parameters are maximized within a prescribed region bounded by process and system constraints, such as maximum torque or power. ACC systems, however, do not use a performance index. In both systems an adaptation *strategy* is used to vary the operating parameters in real time as cutting progresses.

Although there has been considerable research on the development of ACO systems, few, if any, of these systems are used in practice. The major problems with such systems have been difficulties in defining realistic indexes of performance and the lack of suitable sensors which can reliably measure on-line the necessary parameters in a production environment. Practically all the AC systems for cutting processes which are used in production today (1982) are of the ACC type and seldom involve the control of more than one operating parameter.

The objective of most AC systems is improvement in productivity, which is achieved by increasing the metal removal rate (MRR) during rough cutting operations. Several studies have been published [4, 11, 14] which present the productivity increase achieved with AC system as compared to conventional machining (see Fig. 8-2). The increases in productivity range from approximately 20 to 80 percent and clearly depend on the material being machined and the complexity of the part to be produced. The AC systems show the most marked advantages in situations where there are wide variations in the depth of cut during machining as is demonstrated in Fig. 8-2*b*.

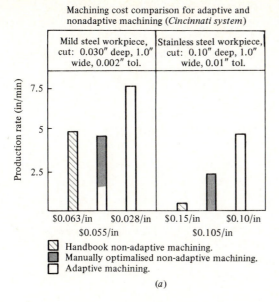

Machining cost comparison for adaptive and nonadaptive machining (*Cincinnati system*)

☑ Handbook non-adaptive machining.
◼ Manually optimalised non-adaptive machining.
☐ Adaptive machining.

(*a*)

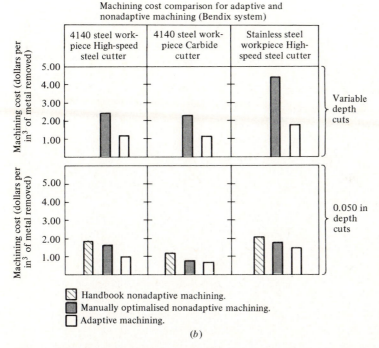

Machining cost comparison for adaptive and nonadaptive machining (Bendix system)

☑ Handbook nonadaptive machining.
◼ Manually optimalised nonadaptive machining.
☐ Adaptive machining.

(*b*)

Figure 8-2 Machining cost comparison for adaptive and nonadaptive machining. (*a*) The Cincinnati Milacron system; (*b*) the Bendix system.

Another saving of time is achieved in the programming stage. Since the operating parameters, such as speeds and feedrates, are adjusted automatically by the system, the part programmer does not have to spend time and effort calculating their optimal values. Again, this time saving becomes significant in complex parts where wide variations in depth of cut must be accommodated.

8-2 ADAPTIVE CONTROL WITH OPTIMIZATION

The best-known research for ACO systems for milling (turning is a similar problem) was conducted at Bendix during the years 1962 through 1964 under the technical supervision of the U. S. Air Force [4]. The block diagram of the Bendix system is shown in Fig. 8-3. The system consists of a Keller-type profiler milling machine, DynaPath NC controller, sensors unit, and adaptive controller. The sensors measure the cutting torque, tool temperature, and machine vibration. These measurements are used by the adaptive controller to obtain the optimal feedrate and spindle speed values as will be later explained.

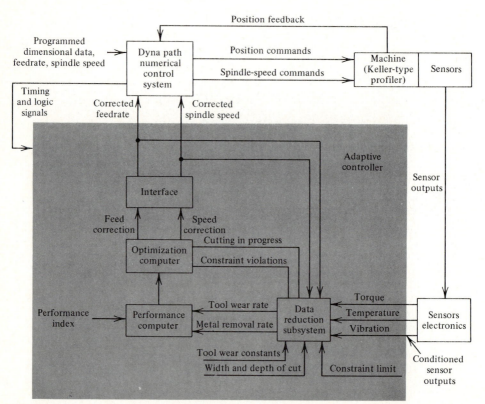

Figure 8-3 ACO system for a milling machine.

The adaptive controller contains two computers and a data reduction subsystem (DRS). Obviously, with today's technology such hardware would be replaced by compatible software in the CNC computer. The DRS is fed by the sensor measurements as well as by the calculated feedrate and spindle speed and a set of constraints. The DRS produces two signals: a metal removal rate (MRR) and a tool wear rate (TWR). The MRR is the product of the milling width (w), the depth of cut (a), and the milling feedrate (V) in inches per minute:

$$MRR = waV \qquad (8\text{-}1)$$

The MRR is used in the calculation of the TWR value:

$$TWR = K_1(MRR) + K_2\theta + K_3\frac{dT}{dt} \qquad (8\text{-}2)$$

where θ is the tool temperature, dT/dt is the time rate of change of the cutting torque, and K_1, K_2, and K_3 are constants which depend on the tool and workpiece material.

The TWR and MRR signals are fed into a performance computer which calculates the performance index ϕ as follows:

$$\phi = \frac{MRR}{C_1 + (C_1t_1 + C_2\beta)(TWR)/W_0} \qquad (8\text{-}3)$$

where C_1 = cost of machine and operator per unit time
C_2 = cost of tool and regrind per change
t_1 = tool changing time
W_0 = terminal allowable width of flank wear
β = adjustable parameter ($0 < \beta < 1$) which determines the type of the performance index (PI):
If $\beta = 1$ the PI is the reciprocal cost per unit.
If $\beta = 0$ the PI is production rate.
If $0 < \beta < 1$ the PI takes into account both the cost per part and the production rate (see Sec. 8-6).

The calculated index ϕ is fed into an optimization computer unit (OCU), which contains the strategy according to which the optimization is performed. The objective of this unit is to continually maintain the value of ϕ at the highest possible value without causing any constraint violations. The constraints in the Bendix system were maximum and minimum spindle speed, maximum torque, maximum feed, maximum temperature, and maximum vibration amplitude. The optimization strategy used by the OCU is based upon the gradient method. With this method a cycle of experimental incremental changes is performed in the feed ($\Delta f = 0.0001$ in/r) and subsequently in the spindle speed ($\Delta N = 10$ rpm). The resultant ϕ is calculated at each step and compared with the previous one. Based upon these calculations the local gradient is determined and the operating point is shifted by a single increment in the direction of this gradient. This cycle is repeated until the maximum value of ϕ is reached.

Although considerable efforts were expended in the Bendix project, it was not commercially accepted. The main problem is that this and other similar ACO systems for milling and turning require on-line measurement of tool wear. So far there have been no industrially acceptable methods developed for the direct measurement of tool wear. Indirect measurement assumes that tool wear is proportional to other measurable variables such as cutting forces and temperatures. However, in addition to tool wear, variations in these parameters can be caused by variations in workpiece hardness and in cutting conditions, thus making it difficult to identify tool wear from variations in these parameters. The Bendix researchers measured the TWR according to Eq. (8-2). However, since the constants K_1 K_2, and K_3 in Eq. (8-2) depend on the tool and workpiece material to use the Bendix system, the user needs to perform off-line experiments to determine the values of these constants for every combination of tool and workpiece material. The time and effort needed for these experiments may override the economic benefits of the AC system.

The lack of a reliable tool wear sensor is the main obstacle in developing industrial ACO systems for milling, drilling, and turning. Although a few techniques were developed to measure the flank wear on turning tools in laboratories, these methods cannot be utilized on the production floor. The problem in milling and drilling is even more difficult because of the complex geometry of milling cutters and twist drills.

Another application of tool wear sensor devices can be for tool change purposes. The tool wear is continuously monitored, and when the tool is almost worn out a tool-change indicator light goes on. This system is sometimes referred to as an "adaptive control system," a term which is a misnomer in describing its task. The tool-change indicator is an open-loop system while a true AC system functions in a closed loop with automatic adjustment of the operating parameters.

8-3 ADAPTIVE CONTROL WITH CONSTRAINTS

We have seen that considerable research and development are required before ACO systems become practical for industrial use. Actually all the AC systems for turning, milling, and drilling used in production today are of the ACC type and seldom involve control of more than one operating parameter. Unlike ACO, ACC systems do not utilize a performance index and are based on maximizing the operating parameters (e.g., feedrate) subject to process and machine constraints (e.g., allowable cutting force on the tool, or maximum power of the machine).

8-3.1 Basic Concepts

The objective of most ACC types of systems is to increase the MRR during rough cutting operations. This is achieved by maximizing one or more operating parameters within a prescribed region bounded by process and system constraints. One useful approach, for example, is to maximize the machining feedrate while maintaining a constant load on the cutter, despite variations in width and depth of cut [2, 10, 16]. This is illustrated in Fig. 8-4 for a slab milling operation. In a normal CNC system, the feedrate is programmed to accommodate the largest width and depth in a particular cut, and this small feedrate is maintained along the entire cut. By contrast, with the ACC system,

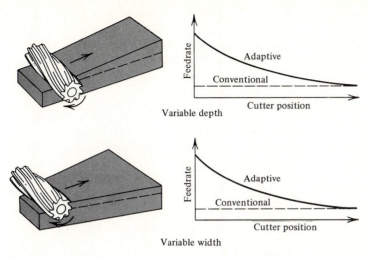

Variable depth

Variable width

Figure 8-4 Automatic compensation for process variables with ACC system.

the maximum allowable load (e.g., cutting force) on the cutter is programmed. As a result, when either the width or depth of cut is increased, the feedrate is automatically reduced, and consequently the allowable load on the cutter is not exceeded. Likewise, when the tool moves through air gaps in the workpiece, the feedrate reaches its maximum allowable value. The ACC system guarantees maximum productivity while minimizing the probability of cutting tool breakage.

ACC systems can also improve the part accuracy by limiting tool deflections due to large cutting forces. This is done by programming a required cutter load which is smaller than the maximum permissable one. This will obviously cause a subsequent decrease in the system productivity. By programming different cutter loads, productivity and accuracy can be traded off in ACC systems.

Commercial AC systems are available for end milling. An end milling cutter may break in bending or in torsion or a tooth may break away. An appropriate ACC system must therefore continuously check the radial cutting force and the cutting torque on the cutter, and vary the feedrate so as to keep both these variables below the permissible limit.

The most commonly used constraints in ACC systems are the cutting force, the machining power, and the cutting torque [6]. The operating parameters are usually the feedrate V (in millimeters per minute or inches per minute) and the spindle speed N (in revolutions per minute); both can be easily manipulated under computer control. The machining feed f is defined by the ratio

$$f = \frac{V}{pN} \tag{8-4}$$

where p is the number of teeth in the cutter in the milling operation; in turning and drilling $p = 1$ is substituted in Eq. (8-4).

The main cutting force F is proportional to the depth of cut a and the feed:

$$F = K_s a f^u \qquad (8\text{-}5)$$

where K_s is the specific cutting force, and u is a parameter in the range $0.6 < u < 1$. Both K_s and u depend on the workpiece and tool material. The cutting torque T is given by the formula

$$T = c_t F D \qquad (8\text{-}6)$$

where D is the workpiece diameter in turning and the tool diameter in milling and drilling. The constant c_t depends on the selected units. If F is given in newtons, T in newton-meters, and D in millimeters, $c_t = 1/2000$. The machining power P is given by the equation

$$P = c_p N T \qquad (8\text{-}7)$$

The constant c_p depends on the selected units. If N is given in revolutions per minute and T in newton-meters, a value of $c_p = \pi/30$ results for the power in watts.

8-3.2 ACC System for Turning

A typical computerized ACC system, which applies the concepts introduced in the previous section, is described for turning on a CNC lathe with a constant cutting force constraint [10].

The ACC system shown in Fig. 8-5 is basically a feedback loop where the feed adapts itself to the actual cutting force and varies according to changes in work conditions as cutting proceeds. The CNC computer executes the original NC control program and an additional AC routine, which is linked to the feedrate routine contained

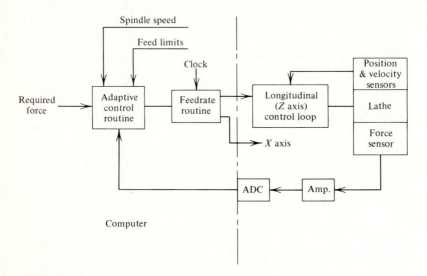

Figure 8-5 ACC system for a lathe.

in the control program. The AC loop functions in a sampled-data mode. The actual main cutting force F is sampled every T seconds (typically $T = 0.1$ s), then converted to a digital signal F_c and sent to the computer. The actual force representation F_c is immediately compared in the computer with a predetermined allowable reference force F_r. The difference between F_r and F_c, which is the force error E ($E = F_r - F_c$) is used as the input to the AC controller. The latter sends a correction signal to the feedrate routine, which, in turn, produces the feedrate command signal. A positive error increases the feedrate and consequently increases the actual force, thereby decreasing the error E, and vice versa.

Controller strategy. The simplest strategy for an AC controller is to provide feedrate corrections *proportional* to the force error E. The basic software structure for such a controller is as follows. The force error is

$$E(i) = F_r(i) - F_c(i) \tag{8-8}$$

where the index (i) indicates the ith sampling interval. The command signal U from the controller is

$$U(i) = U_0 + K_c E(i) \tag{8-9}$$

where K_c is a constant denoted as the controller gain and U_0 is a reference value. The resulting computer feedrate command V_f to the servoloops at the ith sampling (or to the interpolator in a multiaxial mode) is given by

$$V_f(i) = K_f U(i) \tag{8-10}$$

where K_f is a constant associated with the feedrate routine. The feedrate value $K_f U_0$ may be preselected so as to avoid tool breakage when the tool initially impacts the workpiece at the start of the cutting process.

Both K_c and K_f are integral parts of the overall AC open-loop gain K. Due to stability considerations, K should be kept very small (e.g., if $T = 0.1$ s, then $K > 3$ causes instability). However, since the steady-state force error E is inversely proportional to K, it is clear that this error becomes very large for small K's, so that the desired force cannot be achieved. Therefore, the policy given by Eq. (8-9) is not suitable for AC systems for machine tools.

In order to eliminate completely the force error, the controller output command U should be proportional to the time *integral* of the force error W. The simplest structure for such an integral strategy can be written

$$W(i) = W(i - 1) + TE(i) \tag{8-11}$$

and then

$$U(i) = K_c' W(i) \tag{8-12}$$

Equations (8-11) and (8-12) can be combined to give a more efficient form for programming:

$$U(i) = U(i - 1) + K_c E(i) \tag{8-13}$$

where K_c, the controller gain, is proportional to the sampling period T, i.e., $K_c = TK'_c$. As long as there is an error, the command U varies the machine feedrate in a direction to correct this error. At the steady state, however, the error in the force is zero, causing the condition $U(i) = U(i - 1)$, which means that the feedrate command is constant, maintaining the actual force equal to the required one.

This integral strategy has been implemented on a high-power CNC lathe [10]. A typical result for $K_c = 0.5$ is shown in Fig. 8-6. The feed before engagement was selected as 0.5 mm/r. At the start of cutting, the feed is automatically reduced to approximately 0.25 mm/r. The depth of cut is increased by increments of 2 mm, and each time, after a small transient, the force reaches the preselected reference value of $F_r = 1500$ N, and the corresponding feed is decreased.

The selection of the gain K_c is critical to the operation of the AC system. It is known from control theory that the lower the gain K_c, the greater the tendency for stability. Although a small gain causes a sluggish response, the steady-state error always becomes zero.

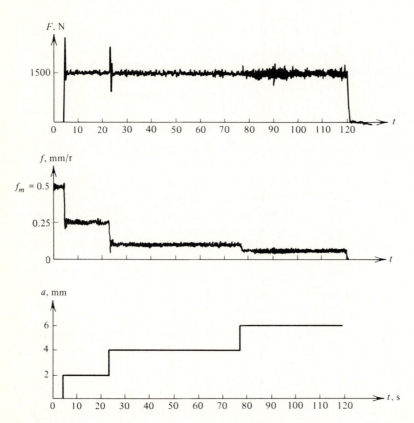

Figure 8-6 The response of the ACC system to changes in depth of cut (gain $K_c = 0.5$; P-25 coated carbide insert; SAE 1045 steel).

8-4 VARIABLE-GAIN AC SYSTEMS

Proper selection of the controller gain K_c is very critical if wide variations in depth of cut, feed, and spindle speed are permitted in the system [10, 12, 14]. The reason is that in the AC system, as is shown in Fig. 8-1, the machining process itself is part of the control loop. Therefore, variations in the process directly affect the control parameters of the loop, and consequently the AC system might become unstable.

8-4.1 The Stability Problem

The results of two turning experiments demonstrating the stability problem are given in Figs. 8-7 and 8-8 [10]. In these experiments the controller gain is $K_c = 0.6$ and the sampling period $T = 0.1$ s. As seen in Fig. 8-7, at the start of cutting the feed is automatically reduced when the depth of cut is increased and the load force on the tool remains constant. However, the system is stable as long as the depth of cut does not exceed 4 mm. At 6 mm, the system becomes unstable with oscillations of approximately 2 Hz. This frequency is in the neighborhood of the natural frequency of the servoloop and is not caused by chatter. Furthermore, when running the system with different spindle speeds and constant depth of cut, the same phenomenon occurs. With a slower spindle speed, the system becomes unstable, as shown in Fig. 8-8.

Instability in AC systems was not a familiar phenomenon to people on the shop floor in the early 1980s, because most of them did not use AC systems in production. Users of AC production systems encountered this instability condition rather infrequently in practice, since their part programmers were experienced enough to avoid large changes in depth of cut and/or spindle speed. This, however, means that the production rate is decreased and that the objective of the AC system is not fully achieved.

The reason for this type of instability prompts a more detailed study of the AC loop. It is known from control theory that the open-loop gain is the dominant parameter in determining system stability. The open-loop gain is equivalent to the sensitivity of the loop at steady state, namely the ratio between the change in the output force F_c to an incremental change in the integrated force error W. The AC open-loop gain is determined as follows.

The relationship between the actual feedrate, or longitudinal axis velocity, V and the command V_f is given at steady state by

$$V = K_n V_f \tag{8-14}$$

where K_n is the gain of the CNC servosystem. The cutting force F is a function of the feed (f) and the depth of cut (a) and can be approximated by Eq. (8-5), which can be written as follows:

$$F = (K_s a f^{u-1})f \tag{8-15}$$

The cutting force F is measured by a force sensor, then converted to a digital word F_c. The conversion factor between F_c and F, including the sensor electronics, is K_e:

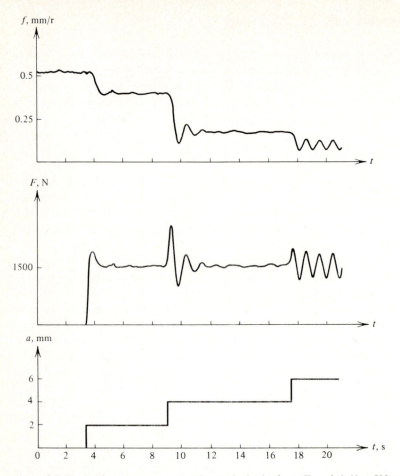

Figure 8-7 The ACC system response to changes in depth of cut (K_c = 0.6; N = 500 rpm).

$$F_c = K_e F \tag{8-16}$$

A steady-state block diagram representation of the whole system is shown in Fig. 8-9. Combining Eq. (8-4) with $p = 1$ and Eqs. (8-10) through (8-16) yields

$$F_c = KW \tag{8-17}$$

where K, the AC open-loop gain, is defined by

$$K = K'_c K_f K_n K_e K_s \left(\frac{a}{N}\right) f^{u-1} \tag{8-18}$$

and has the dimensions of s^{-1}.

It is known from control theory that most systems have an upper allowable limit to their open-loop gain. This limit is prescribed by the dynamics of the system and the sampling period T and if exceeded the system becomes unstable. A satisfactory gain is

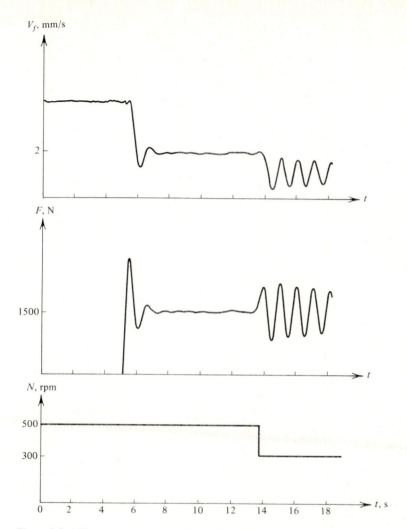

Figure 8-8 ACC system response to changes in spindle speed ($K_c = 0.6$; $a = 3$ mm).

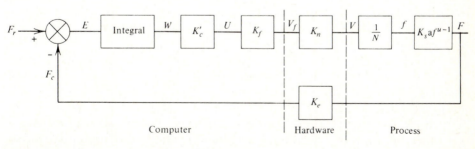

Figure 8-9 Block diagram of the ACC system.

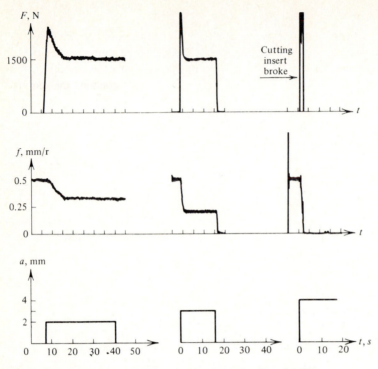

Figure 8-10 Tool breakage due to controller low gain ($K_c = 0.0625$).

usually regarded as about one-half of this value. If K is much smaller, the transient response is very slow.

Since the depth of cut and the spindle speed are contained in K, an increase of the first, or a decrease of the latter, can cause instability conditions, as seen in Figs. 8-7 and 8-8. One might think that a possible solution is the selection of a very small K_c in Eq. (8-13) to decrease the open-loop gain under its stability limit even at the largest allowable depth of cut and a minimum permissible spindle speed. The result of this approach is demonstrated in Fig. 8-10 for $K_c = 0.0625$. At a depth of cut of 2 mm the transient behavior is very slow and the steady-state force reaches the preselected reference value of $F = 1500N$ only after a relatively long time. But if the chip load is too big ($a = 4$ mm), the recovery time from the initial impact is too long and the tool insert breaks.

We see that the selection of K_c is critical to the performance of the AC system. If K_c is too large, the entire CNC-AC system can become unstable. When K_c is too small, the transient behavior is very sluggish and, as a result, the tool insert may break at medium to large depths of cut. This calls for a different approach to AC system design. The system should operate with a variable-gain K_c which adapts itself to the cutting parameters. This involves an estimation in real time of the gain of the cutting process, and a subsequent adaptation of the AC controller gain (K_c) to the changing conditions of the cutting process.

8-4.2 The Estimator Algorithm

The cutting process estimator should measure in real time the quantity $K_s \, (a/N)f^{u-1}$ which affects the open-loop gain in Eq. (8-18). However, since a direct estimation of this quantity requires additional sensors and output channels to the computer, it is worthwhile to estimate the value of a process gain K_p, which contains the required quantity and is defined by

$$K_p = K_f K_n K_e K_s \left(\frac{a}{N}\right)f^{u-1} \tag{8-19}$$

By definition, at steady state K_p is given by

$$K_p = \frac{F_c}{U} \tag{8-20}$$

Since the values of both F_c and U are available within the computer, the process gain can be calculated. Subsequently, the controller gain K_c (where $K_c = TK_c'$) could be adjusted in real time according to the equation

$$K_c = \frac{TK}{K_p} \tag{8-21}$$

where K is the desired open-loop gain.

The subroutines which perform the process estimation and the AC controller gain adjustment should function in real time and be programmed on the CNC computer, which is typically a simple microcomputer. Therefore, it is recommended that these subroutines be written in assembly language using only basic instructions such as add, subtract, compare, etc. Unfortunately, implementation of the algorithm based upon Eqs. (8-20) and (8-21) requires a direct division, which is a complicated and slow operation in assembly language.

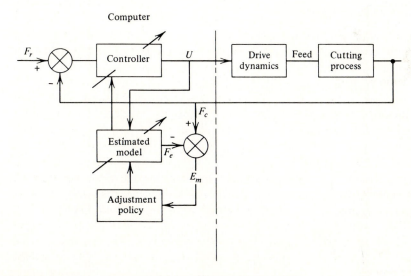

Figure 8-11 Process estimation in a variable-gain ACC system.

A block diagram of an alternative approach which avoids direct division is shown in Fig. 8-11. The estimated model block contains an estimated gain K_m, which is multiplied by the input U to generate an estimated force F_e. In general, $K_m \neq K_p$, and an error E_m is generated:

$$E_m = F_c - F_e = F_c - UK_m \tag{8-22}$$

Since at steady state $F_c = UK_p$, the model estimation error is

$$E_m = U(K_p - K_m) \tag{8-23}$$

The estimated model gain K_m should be automatically adjusted to reduce this error. The simplest adjustment policy which guarantees a zero error is to apply an integration algorithm:

$$K_m = C \int E_m \, dt \tag{8-24}$$

With this algorithm when the error E_m is zero, K_m is a constant which satisfies $K_m = K_p$. In the computer program this estimator algorithm is given by the following equations:

$$E_m(i + 1) = F_c(i) - U(i)K_m(i) \tag{8-25}$$

$$K_m(i + 1) = K_m(i) + K_1 E_m(i + 1) \tag{8-26}$$

which can be readily implemented in assembly language. The estimator gain K_1 can be selected to be 2^{-n}, which simplifies the multiplication in Eq. (8-26) to n shift operations. The specific value of K_1 depends on the amount of noise in the measured force. In the presence of high-level noise, the estimator gain must be small in order to smooth the noise and estimate the process parameter K_m with good precision. This however causes the disadvantage of slower convergence to the steady-state value of K_m.

8-4.3 Variable-Gain Algorithm

The objective of the proposed adaptive control loop is to maintain a constant open-loop gain despite variations in the cutting parameters. By combining Eqs. (8-18) and (8-19) the open-loop gain is defined by

$$K = K_c' K_p \tag{8-27}$$

and substitution of the estimated value K_m for K_p yields

$$K = K_c' K_m = \frac{K_c K_m}{T} \tag{8-28}$$

The constant open-loop gain can be obtained by adjusting the controller gain K_c according to variations of K_m. As in the case of the estimation algorithm, direct division is avoided and smoothing is achieved by using the following integration policy:

$$E_c(i + 1) = TK - K_c(i)K_m(i) \tag{8-29}$$

$$K_c(i + 1) = K_c(i) + K_2 E_c(i + 1) \tag{8-30}$$

Again, the integration algorithm in Eq. (8-30) guarantees that $E_c = 0$ at the steady state, which means that the desired gain can be achieved.

We see that the solution to the original problem is in the form of a supplementary adaptation loop. In the main loop the machining feed is adapted to maintain a constant load on the cutting tool, and in the supplementary loop the controller gain is adapted to maintain a constant loop gain despite variations in the cutting conditions. This

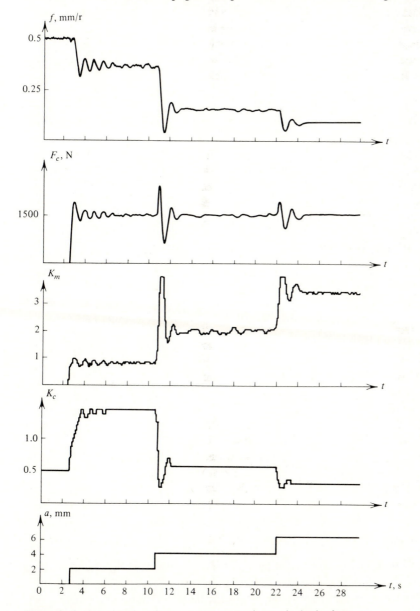

Figure 8-12 Variable-gain AC system response to changes in depth of cut.

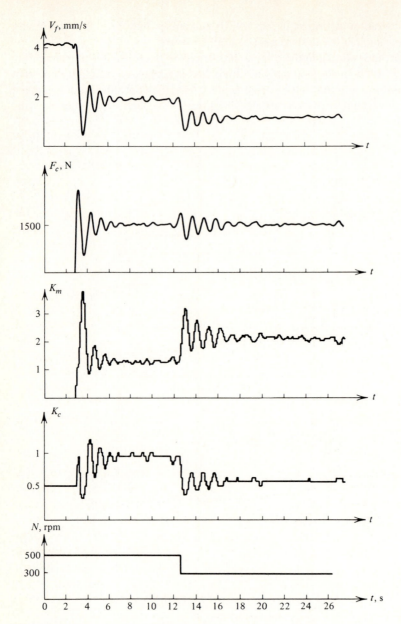

Figure 8-13 Variable-gain AC system response to changes in spindle speed.

technique is also known as a model reference adaptive system (MRAS), in which a reference model appears as part of the control system [9].

A set of experiments was performed to compare the performance of a conventional AC system with the variable-gain AC system [8]. In these experiments the controller gain in the conventional AC was $K_c = 0.6$, the sampling period $T = 0.1$ s, and the integration constants in the variable-gain AC system were set to $K_1 = K_2 = \frac{1}{4}$. The AC open-loop gain has an upper limit K_ℓ obtained from stability considerations. An acceptable K should be about one-half of this limiting value. However, in order to demonstrate the performance of the variable-gain system, an open-loop gain of $0.9K_\ell$, which is very close to the stability limit, was selected.

The objective of the experiments shown in Figs. 8-12 and 8-13 was to remedy the unstable conditions obtained in Figs. 8-7 and 8-8 for the conventional AC system. In Fig. 8-7, it can be seen that the conventional AC system became unstable for a depth of cut of 6 mm. By contrast, as is seen in Fig. 8-12, the variable-gain AC system was always stable. The value of K_c was automatically reduced from 1.5 to 0.52 and then to 0.33 with the progressive increase in the depth of cut from 2 to 4 and then to 6 mm. Likewise, in Fig. 8-8 the conventional AC became unstable with the decrease of spindle speed from 500 to 300 rpm. By contrast, the variable-gain system (Fig. 8-13) adapted itself automatically to the speed change and the system remained stable by an automatic reduction in the gain K_c. The oscillations at the transient periods can be eliminated by reducing the open-loop gain to a lower value, e.g., $K = 0.5K_\ell$.

The variable-gain AC system provides a solution to the stability problem which arises in conventional ACC systems when wide variations in depth of cut are needed. The solution does not require any additional hardware, and all the modifications to the AC controller are implemented by software. Similar variable-gain loops can be applied to other manufacturing systems, e.g., to a robot arm. In this case the loop gain should be adapted to the changes in moment of inertia caused by different loads and by changing of the arm position during the motion of the robot.

8-5 ADAPTIVE CONTROL OF GRINDING

Although there has been considerable research on ACO systems for milling, turning, and drilling, none of these systems is used in practice. The main difficulties encountered with such systems have been mainly in specifying an index of performance together with an appropriate control policy which is not too complex, and in developing sensors which can reliably measure the necessary process variables in a production environment. One process for which ACO machine tool systems appear to be promising is grinding. One prototype system, a computerized ACO system for plunge grinding of steels [1], is described below. The objective of this system is to maximize removal rate, subject to constraints on workpiece burn and surface finish, by controlling the grinding parameters based upon on-line sensing of the grinding power.

8-5.1 Grinding Model

The optimization strategy for the ACO system is based mainly upon Malkin's grinding model for plunge grinding [1]. The essential aspects of this model, which are summarized in this section, include the partition of the grinding power among its fundamental components and the prediction of the critical grinding power for the burning constraint. The full model includes also the dependence of surface finish on process parameters, and the influence of dressing on grinding performance [1].

The total grinding power can be considered to consist of chip formation, plowing, and sliding components:

$$P = P_c + P_p + P_s \tag{8-31}$$

Based on experimental studies each power component can be expressed in terms of the operating parameters as follows:

$$P_c = K_c(\pi d_w v_f w) \tag{8-32}$$

$$P_p = K_p w v_s \tag{8-33}$$

and
$$P_s = \left(K_{s1} + K_{s2}\frac{\pi n_w d_w}{v_s d_e} \right) d_e^{1/2} v_f^{1/2} n_w^{-1/2} A \tag{8-34}$$

where d_w is the workpiece diameter (in millimeters), w is the grinding width (in millimeters), v_f is the radial infeed velocity (in millimeters per second), v_s is the peripheral wheel velocity (in meters per second), n_w is the rotational speed of the workpiece (in revolutions per second), and d_e is the equivalent diameter (in millimeters) which for external cylindrical grinding is given by

$$d_e = \frac{d_s d_w}{d_s + d_w}$$

where d_s is the wheel diameter. The parameter A represents the effective "wear flat area"† on the grinding wheel.

One main limitation on the removal rate for grinding of steels is workpiece burn. The corresponding threshold grinding power to reach a critical grinding zone temperature for workpiece burn can be written in terms of the grinding parameters as

$$P_b = K_{b1} w d_w v_f + K_{b2} w d_e^{1/4} v_f^{1/4} n_w^{1/4} d_w^{1/2} \tag{8-35}$$

where K_c, K_p, K_{s1}, K_{s2}, K_{b1}, and K_{b2} in Eqs. (8-32) through (8-35) are grinding constants. The power and the burning equations are programmed in the computer and used by the adaptive controller.

8-5.2 Optimization Strategy

The performance index of the grinding process is the volumetric removal rate which for external cylindrical grinding can be written

$$\text{MRR} = \pi d_w v_f w \tag{8-36}$$

The objective of the optimization is to maximize MRR subject to the constraints of workpiece burn and surface finish requirements. For any particular optimization, the workpiece diameter d_w is nearly constant and the grinding width w is fixed, so the performance index can be simply taken as the radial infeed velocity v_f.

---

† A measurement of the dullness of a grinding wheel.

This text discusses a simplified optimization problem with fixed dressing where the objective is to maximize v_f subject only to the burning constraint without any surface finish constraint. In this case, the problem can be written:

Maximize: v_f

Subject to: $P = P_b$

(8-37)

From Eqs. (8-31) to (8-35), it can be shown [1] that for a specified wear flat area A there is a particular combination of v_f and n_w which will satisfy Eq. (8-37), and this is the optimal working point (v_f^*, n_w^*) for a given equivalent diameter, wheel diameter, and wheel speed. The collection of all the optimal points for various wear flat areas defines an optimal locus in the $v_f n_w$ plane as shown in Fig. 8-14. Any point on an optimal locus is the optimal operating point for a particular wear flat area; optimal points further out along the locus at larger removal rates correspond to sharper wheels.

The optimal locus provides a convergence path to the optimal working point provided that the grinding power can be measured on line. At the optimal working point, the equality $P = P_b$ prevails. Starting from some initial point on the optimal locus where $P < P_b$, the optimal working point can be reached by proceeding out along the optimal locus to faster infeed velocities until the measured power is equal to the corresponding burning power given by Eq. (8-35). Likewise, if $P > P_b$ the optimal working point can be reached by backing off along the optimal locus until $P = P_b$.

In addition to the surface finish and burning constraints, other constraints can arise due to machine tool limitations, as illustrated schematically in Fig. 8-14, together with the optimal locus. These constraints include an upper limit on the infeed velocity (v_{fx}), upper and lower limits on the workpiece spindle speed (n_{wx} and n_{wm}), and maximum available grinding power (P_x). Taking these additional factors into account, the same optimization strategy described above still applies, provided that a modified optimal locus ABCDE in Fig. 8-14 is used.

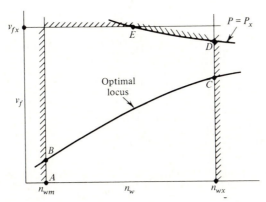

Figure 8-14 Illustration of optimal locus in grinding, together with machine constraints.

8-5.3 Design of Adaptive Control for Grinding

The overall concept of the ACO system is shown in Fig. 8-15 which includes a grinding machine interfaced with a computer. The two controlled grinding parameters are the workpiece spindle speed n_w and the radial infeed velocity v_f. Measurements of the grinding power P are fed from the grinding machine to the computer, and the computer in turn controls the parameters v_f and n_w to operate along the optimal locus (or modified optimal locus) and converge toward the optimal working point. The workpiece spindle is driven by a dc servomotor in the continuous range 0.4 to 63 r/s ($n_{wm} = 0.4$ r/s and $n_{wx} = 63$ r/s). For controlling the radial infeed velocity v_f, a stepping motor drive was attached to the infeed control handwheel of the machine. A Hall element sensor was connected to the main drive motor to measure the machine power. During the grinding operation, the on-line measured power is directly fed into the computer, and the net grinding power P is obtained by subtracting the idling power from the measured power.

The adaptive control system is of the sampled-data type. Every time period T the measured power P is sampled; the burning power P_b is calculated using the previous v_f and n_w values. The new control parameters v_f and n_w are then assigned. Between sampling events, the values of the control parameters are kept constant by storing their values in computer registers assigned to DACs.

The heart of the control system is an algorithm incorporating the optimal locus optimization strategy described in the previous section. The grinding operation might start at an arbitrary point in the $v_f n_w$ plane but is immediately transferred by the computer to a point on the optimal locus by changing n_w. Thereafter, the trajectory of convergence toward the optimal operating point is along the optimal locus. For controlling the rate of convergence, instead of providing an external reference as is typically done, the reference to the control loop P_b is calculated according to Eq. (8-35) in the block G_b to which the control variables v_f and n_w are fed. As a consequence, the convergence rate depends on the error $e = P_b - P$, which converges to zero when

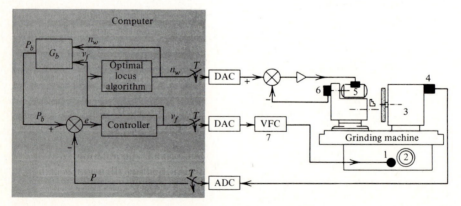

Figure 8-15 On-line adaptive control system for grinding. 1-Stepping motor infeed drive; 2-infeed control handwheel; 3-grinding wheel motor; 4-power sensor; 5-workpiece spindle dc motor; 6-tachometer; 7-voltage-to-frequency converter (VFC).

proceeding along the locus. It should be noted that if the available grinding power P_x is less than the allowed power P_b, the value of P_x is used in place of P_b when calculating the error e.

From preliminary grinding experiments, the grinding power P was found to be much more sensitive to infeed velocity v_f than to the spindle speed n_w. Therefore, only v_f is directly determined by the controller in Fig. 8-15 and the corresponding n_w is calculated on the optimal locus. Such a single-input–single-output controller is much simpler to design than a multioutput controller. The controller algorithm was selected according to the equation:

$$v_f(i) = v_f(i-1) + K_1 e(i) + K_2 [e(i) - e(i-1)] \tag{8-38}$$

where the index i is the number of the sampling event. The first two terms in Eq. (8-38) constitute an integral controller of gain K_1 which ensures a zero error ($e = 0$) at the optimal point. The last term is essentially a derivative controller of gain K_2 which was added to decrease the tendency for overshooting while converging toward the optimal point. The controller gains of the pilot system were selected as $K_1 = 3.3 \ \mu\text{m}/(\text{s} \cdot \text{kW})$ and $K_2 = 5.3 \ \mu\text{m}/(\text{s} \cdot \text{kW})$. These controller gains were used except when the operation approached close to the optimal working point, as defined in the present case as the error being within 5 percent of the allowable power. At this point, a fast response is no longer needed to maintain optimal grinding conditions, so the gains K_1 and K_2 are substantially reduced and a software low-pass filter is added to the measured power line.

For repetitive grinding of identical parts (referred to as cycles), the AC system makes use of a learning starting point. After the first part is ground, the system selects from the optimal grinding conditions obtained with that part as a starting point for the next part. This starting point could be the same as the optimum working point reached at the end of the previous part, or it might be preferable to choose a slightly lower infeed rate if there is significant part-to-part grinding variability.

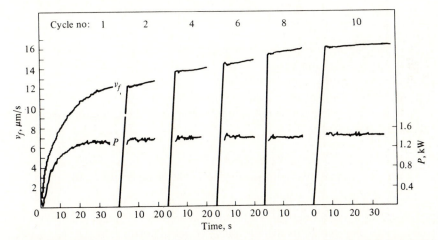

Figure 8-16 Grinding of repetitive cycles with power limit constraint: 32A46K8VBE wheel, SAF 4340 steel workpiece (240 Brinell), $d_s = 440$ mm, d_w 165 mm, $w = 32$ mm; 0.3 mm removed from workpiece radius per cycle.

Some results illustrating operation of the grinding system for repetitive grinding cycles with only the grinding power constraint are presented in Fig. 8-16. For cycle 1, it can be seen when starting from an initial infeed velocity $v_f = 1 \mu$m/s that it takes 28 s for the process to converge near to the optimal point where the power line filter is activated and the gains reduced, as seen from the smoothing of the v_f curve. In the subsequent cycles, the system accelerated to the learning starting point, and the convergence time was reduced to about 4 s. For this series of experiments, the grinding wheel was dressed only before the first cycle. The increase in optimum infeed velocity which can be seen from one cycle to the next indicates self-sharpening of the wheel (wear flat area decreases) which is accompanied by a deterioration of surface finish. For redressing of the wheel, say after cycle 10, the learning starting point for the subsequent cycle could be selected on the basis of the optimal operating condition reached in cycle 1.

This and other computerized ACO grinding systems which are designed to maximize removal rate can be implemented industrially. An existing CNC grinder can be upgraded to an ACO system with only the addition of a power or force sensor and some additional software incorporating the optimal control strategy.

8-6 COST ANALYSIS IN MACHINING

Determination of the optimal cutting speed through cost analysis is somewhat related to adaptive control. The fundamental difference is, however, that while this optimization is an off-line procedure, adaptive controls always function in on-line.

The total cost for any machining operation is comprised of four individual costs:

1. Actual part cutting cost $= C_1 t$
2. Tool cost (including grinding cost) $= C_2(t/T)$
3. Tool changing cost $= C_1 t_1 (t/T)$
4. Loading and unloading cost $= C_1 t_2$

where C_1 = direct operating cost, which includes machine and labor (direct + overhead) cost per unit time ($/min)

C_2 = tool cost, which can be either dollars per cutting edge in throwaway tools or original cost of cutter plus regrinding cost

t = the time that the tool is removing metal (direct machining time per part)

t_1 = tool changing time

t_2 = handling time

T = the tool life, namely the number of minutes that a cutting tool can cut before replacing the cutting edge or between successive tool regrinding operations.

The term t/T in the tool cost and tool changing components is the number of tool edges required per workpiece, or the reciprocal of the number of parts that can be machined with one cutting edge (or for each tool sharpening).

The loading and unloading cost is not affected by the operating parameters, such as cutting speed, and apparently cannot be optimized by their manipulations. The remainder of this text is concerned with the determination of the optimum machining parameters and therefore this cost component is not considered.

Hence, the unit cost ψ is the sum of the other three cost components:

$$\psi = C_1 t + (C_2 + C_1 t_1)\frac{t}{T} \tag{8-39}$$

The direct machining time t can be defined as the ratio between the removed volume Q and the MRR, which is expressed as cubic millimeters (or inches) per minute:

$$t = \frac{Q}{MRR} \tag{8-40}$$

On a lathe, for turning operations, the MRR is obtained from the expression $1000vfa$ mm^3/min or $12vfa$ in^3/min, where v is the cutting speed (in meters or feet per minute), f is the feed (in millimeters or inches per revolution), and a is the depth of cut (in millimeters or inches). The removed volume is given by the expression $Q = a\pi DL$, where D is the diameter of the workpiece and L is the axial length of cut. Likewise in milling the MRR is given by Eq. (8-1), and the removed volume does not depend on v and f. Therefore, the time t in Eq. (8-40) is inversely proportional to v and f.

In most cost analysis situations the feed and depth of cut are fixed by machine limitations and functional specifications of the part being machined. Even when the feed is an admissible variable, there are no optimum values for both feed and speed, and it is more efficient to remove metal with the highest allowable feed. Consequently Eq. (8-40) may be simplified to

$$t = \frac{C_3}{v} \tag{8-41}$$

The direct machining time t may be reduced if the metal is removed at a faster rate. Increasing the cutting speed, for example, will decrease the time t in *direct proportion* and the first term in Eq. (8-39) will be decreased as well. However, with the increase in cutting speed the tool life is *exponentially* decreased and therefore the second term in Eq. (8-39) is increased. As a consequence an optimum cost according to Eq. (8-39) exists only if the tool life decreases exponentially with a linear increase in an operating parameter which directly affects the machining time t.

For many machining operations the tool life is given by Taylor's tool life equation:

$$vT^n = C \tag{8-42}$$

where C and n are empirical constants which depend on the workpiece and tool materials, and, in addition, C depends also on the feed, depth of cut, cutting fluid, etc. The constant C is the cutting speed that results in a 1-min tool life. For example, if $C = 100$ m/min, the tool will fail in 1 min if the machining is performed at a cutting speed of 100 m/min. The exponent n varies from 0.05 to 0.15 for most work materials when a high-speed steel (HSS) tool is used; from 0.2 to 0.35 for carbide tools; and from 0.35 to 0.5 for ceramic tools. As long as the condition $n < 1$ is satisfied, an optimal cutting speed which minimizes the cost ψ can be determined.

The cutting speed for a *minimum cost* can be derived by substituting Eqs. (8-41) and (8-42) into Eq. (8-39), differentiating the cost with respect to cutting speed and setting the result equal to zero, which yields

$$v_m = \frac{C}{[(1/n - 1)\tau]^n} \tag{8-43}$$

where

$$\tau = t_1 + \frac{C_2}{C_1}$$

This optimal speed is calculated before the machining operation. The corresponding minimum tool life T_m is

$$T_m = (\frac{1}{n} - 1)\tau \tag{8-44}$$

If the desired criterion is *maximum production rate* rather than minimum cost, then the tool cost C_2 is insignificant and the criterion becomes $1/\psi$ with the substitution of $C_2 = 0$. The optimal speed for maximum production is also given in Eq. (8-43), but with the substitution $\tau = t_1$. As a consequence this speed is larger than the one associated with minimum cost. If the *maximum profit* is the desired criterion, the optimal speed lies between the other two optimal speeds [20].

In ACO systems the operating parameters are adjusted in real time and consequently the tool life must be predicted during the machining itself. A simple method may be by using the ratio

$$T = \frac{W_0}{\text{TWR}} \tag{8-45}$$

where W_0 is the final tool wear at $t = T$ and TWR is the instantaneous tool wear rate. Equation (8-45) was used by the Bendix researchers in the development of their ACO system [4]. They used a performance index defined by

$$\phi = \frac{Q}{\psi} \tag{8-46}$$

Substituting Eqs. (8-39), (8-40) and (8-45) into Eq. (8-46) results in

$$\phi = \frac{\text{MRR}}{C_1 + (C_2 + C_1 t_1)(\text{TWR})/W_0} \tag{8-47}$$

which is similar to Eq. (8-3) with $\beta = 1$. By adding the adjustable parameter β, the performance index can be maximum production rate ($\beta = 0$), maximum profit ($0 < \beta < 1$), or minimum cost ($\beta = 1$).

BIBLIOGRAPHY

1. Amitai, G., S. Malkin, and Y. Koren: Adaptive Control Optimization of Grinding, *Trans. ASME, J. Eng. Ind.,* vol. 103, pp. 102–111, February, 1981.
2. Beadle, B. R., and J. G. Bollinger: Computer Adaptive Control of Machine Tools, *CIRP Ann.,* vol. 19, 1970.
3. Bedini, R., G. G. Linsini, and P. C. Pinotti: Experiments on Adaptive Control of a Milling Machine, *Trans. ASME, J. Eng. Ind.,* vol. 98, pp. 239–245, February, 1976.
4. Centner, R.: Final Report on Development of Adaptive Control Technique for Numerically Controlled Milling Machine, *USAF Tech. Documentary Report ML-TDR-64-279,* August, 1964.
5. Colwell, L.W., J. R. Frederick, and L. J. Quackenbush: "Research in Support of Numerical and Adaptive Control in Manufacturing," The University of Michigan, Ann Arbor, 1969.
6. Cook, N. H., K. Subramanian, and S. A. Basile: "Survey of the State of the Art of Tool Wear Sensing Techniques," Massachusetts Institute of Technology, September, 1975.
7. Inamura, T., T. Senda, and T. Sata: Computer Control of Chattering in Turning Operation, *CIRP Ann.,* vol. 26, pp. 181–186, 1977.
8. Koren, Y., and O. Masory: Adaptive Control with Process Estimation, *CIRP Ann.,* vol. 30, no. 1, pp. 373–376, 1981.
9. Landau, I. D.: "Adaptive Control," Academic Press, New York, 1979.
10. Masory, O., and Y. Koren: Adaptive Control System for Turning, *CIRP Ann.,* vol. 29, no. 1, pp. 281–284, 1980.
11. Mathias, A.: An Effective System for Adaptive Control of the Milling Process, *SME Tech. Paper MS-68-202,* 1968.
12. ————: An Adaptive Controlled Milling Machine, *SME Tech. Paper MS-76-260,* 1976.
13. ————: Software Adaptive Control—Optimum Productivity for CAM, *SME Tech. Paper MS-77-252,* 1977.
14. Porter, R. D., and M. J. Summers: Adaptive Machine Tool Control, the State of the Art, *Mach. Prod. Eng.,* pp. 214–220, February 5, 1969.
15. Stute, G.: Adaptive Control, *Proc. Mach. Tool Task Force Conf.,* vol. 4, chap. 7–14, October, 1980.
16. Tlusty, J., Y. Koren, and P. Macniel: Numerical and Adaptive Control for Die Sinking, *Proc. Intern. Conf. Prod. Eng.,* Tokyo, 1974.
17. ————, and M. Elbestawi: Analysis of Transient in an Adaptive Control Servo Mechanism for Milling with Constant Force, *Trans. ASME, J. Eng. Ind.,* vol. 99, no. 3, pp. 766–772, August, 1977.
18. ————, and M. Elbestawi: Constraints in Adaptive Control with Flexible End Mills, *CIRP Ann.,* vol. 28, pp. 253–255, 1979.
19. Weck, M., E. Verhaag, and M. Gather: Adaptive Control for Face-Milling Operations with Strategies for Avoiding Chatter-Vibrations and for Automatic Cut Distribution, *CIRP Ann.,* vol. 24, pp. 405–409, 1975.
20. Wu, S. M., and D. S. Ermer: Maximum Profit as the Criterion in the Determination of the Optimum utting Conditions, *Trans. ASME, J. Eng. Ind.,* vol. 88, no. 4, November, 1966.

PROBLEMS

8-1 A milling cut (250 mm long) is taken across the workpiece shown in Fig. 8-4, in which the depth of cut a is linearly changed from 2.5 to 5 mm along the cut (constant width). The programmer of a CNC system programmed the part with a constant feedrate of $V = 2$ mm/s according to the worst-case condition (i.e., $a = 5$ mm). Calculate the machining time.

8-2 The part of Prob. 8-1 is machined using an adaptive control system (ACC type). Assume that for a constant spindle speed, the condition $aV = 10$ maintains a constant load on the cutting tool. The ACC system would vary the feedrate along the cut to maintain this condition. Calculate the feedrates corresponding to $a = 2.5, 3, 3.5, 4,$ and 5 mm and draw the feedrate versus milling length.

8-3 Determine the machining time with the ACC system in Prob. 8-2, and compare it with the one achieved with conventional CNC in Prob. 8-1.

8-4 Machinability tests to determine the optimal cutting speed were conducted on a lathe. The Taylor equation obtained for throwaway carbide tools and a steel workpiece was $vT^{0.3} = 100$, where v is in meters per minute and T in minutes. The cost of each tool is \$1.20 (for 6 cutting edges), the machine and operator rate is \$40/h, and the tool changing time is 0.5 min. Calculate

(a) The cutting speed for the minimum cost.
(b) The tool life that will result in the maximum production rate.
(c) The minimum cost for one turning operation ($a = 2$ mm, $f = 0.25$ mm/r, $L = 250$ mm, $D = 51$ mm).
(d) The cost when turning at the maximum production-rate speed.
(e) The difference in cost when cutting with a speed 10 percent lower than the optimal.

NINE

INDUSTRIAL ROBOTS

Industrial robots are beginning now to revolutionize industry. These robots do not look or behave like human beings, but they do the work of humans. Robots are particularly useful in a wide variety of applications, such as material handling, spray painting, spot welding, arc welding, inspection, and assembly. Current research efforts focus on creating a "smart" robot that can "see," "hear," and "touch" and consequently make decisions.

The technology of robots is related to, but differs somewhat from, NC technology in that robots effect higher velocity and movement in more axes of motions. While with NC only a *point,* namely the endpoint of the cutter, is controlled in the space, with robots both the *endpoint* and *orientation* are manipulated. This requires more degrees of freedom, more powerful software, and more effective control algorithms.

9-1 BASIC CONCEPTS IN ROBOTICS

The industrial robot is a programmable mechanical manipulator, capable of moving along several directions, equipped at its end with a device called the end-effector, and performs factory work ordinarily done by human beings. The term robot is used for a manipulator that has a built-in control system and is capable of stand-alone operation.

The *Webster's* dictionary[†] defines a robot as "any mechanical device operated automatically to perform in a seemingly human way." By this definition, a garage door opener, which automatically opens the door by remote control is also a robot. Obviously this is not an industrial robot. The Robot Institute of America (RIA) defines the industrial robot as "a reprogrammable multi-functional manipulator designed to move

† Second college edition, 1980.

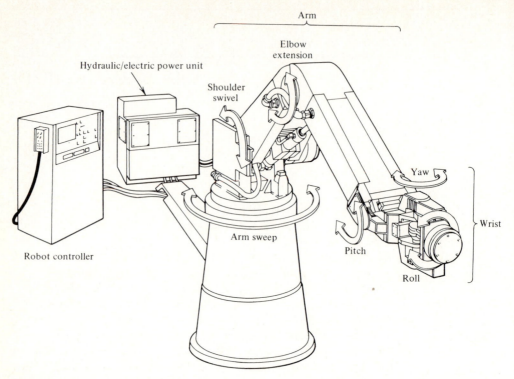

Figure 9-1 Structure of industrial robot system. (*Courtesy of Cincinnati Milacron.*)

material, parts, tools, or other specialized devices through variable programmed motions for the performance of a variety of tasks." By this definition, however, a washing machine is also a robot; the wash and rinse cycles are programmable and the machine moves material in rotary motions. Therefore, a definition of an industrial robot must include the following key words: programmable manipulator, end-effector, factory work, and stand-alone operation. If one or more of these key words is missing, then washing machines, traffic lights, special purpose mechanisms and manufacturing machines for mass production are defined as robots as well, and this is not our intention.

In general, the structure of a robot manipulator is composed of a *main frame* and a *wrist* at its end. The main frame is frequently denoted as the *arm*, and the most distal group of joints affecting the motion of the end-effector is referred to collectively as the *wrist*. This typical structure is shown in Fig. 9-1. The end-effector can be a welding head, a spray gun, a machining tool, or a gripper containing on-off jaws, depending upon the specific application of the robot. Each of these devices is mounted at the end of the robot and performs the work, and therefore is also denoted as the *robot tool*.

Basically the robot needs *six* axes of motion (or degrees of freedom) to reach a *point* with a specific *orientation* in the space. A different orientation might completely change the position of the robot arm. For example, to place a weld on the top side of

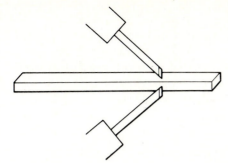

Figure 9-2 Different orientations in beam welding.

the beam in Fig. 9-2 requires a completely different orientation from that required to place a weld at almost the same point but on the bottom side of this beam. Typically the arm has three degrees of freedom, in linear or rotary motions, and the wrist section contains three rotary motions. The combination of these six motions will orient the welding head, or other tool, and position it at the required point in the space.

Each axis of motion is separately driven by an actuator which may be a dc servomotor, a stepping motor, a pneumatic actuator, or a hydraulic motor or actuator. The drive elements may be coupled directly to the mechanical links or joints, or may drive them indirectly through gears, chains, cables, or leadscrews.

The first-generation[†] robots have relatively primitive control strategies and are often referred to as "pick and place" robots or "non-servo" robots. The drive elements of these robots are typically pneumatic or hydraulic cylinders. The feedback devices are simply pairs of limit switches and stoppers for each axis of motion of the arm. The principle of operation of these robots is similar to that of automatic lathes with sequential control as described in Chap. 1. At each step the controller of the robot sends a control signal to the valve of a desired axis to move it. The motion continues until restrained by the stopper and its corresponding limit switch. The limit switch sends a signal to the controller, which then commands the valve to close and proceeds to the next step in the control sequence involving another valve. This process is repeated until the entire sequence of steps has been executed.

Although the pick and place robots are relatively inexpensive, they have a very low control flexibility, since the number of motions in a program is limited by the number of limit switches that can be installed. Further discussion will focus on the more advanced robots that have a high programming and control flexibility. These robots are often referred to as the second-generation robots, or servo-controlled robots, because they are generally computerized devices which are controlled in closed-loops by servo-drives. The third-generation robots are capable of making decisions and generating unprogrammed motions based upon the information sent by sensors, such as a TV camera for vision, or a force and pressure transducer for force feedback.

---

† Generation means in this text a stage of improvement in the development of a system.

9-2 THE MANIPULATOR

The manipulator is the mechanical unit which performs the movement function in the robot. It consists of a series of mechanical links and joints capable of producing controlled movement in various directions. The manipulator is composed of the main frame (the arm) and the wrist; each has three degrees of freedom, or axes of motion. Structurally the robots can be classified according to the coordinate system of the main frame:

Cartesian—three linear axes
Cylindrical—two linear and one rotary axis
Spherical—one linear and two rotary axes
Articulated, or jointed—three rotary axes

In order to aid in defining the coordinate system, a symbolic notation is often used to describe the types and number of joints, starting from the base to the end of the arm. Linear joints can be sliding or prismatic, designated S or P, and revolute joints designated R. By this notation the spherical robotic arm, for example, would be called an RRP arm, and the articulated one the RRR arm.

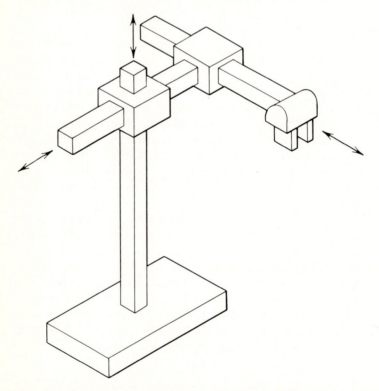

Figure 9-3 Cartesian coordinate manipulator. (*Courtesy of U.S. Air Force.*)

One of the most important characteristics in robotics is the shape of the arm reach envelope, or the arm *work volume*. The shape of the work volume depends on the coordinate system, and its size depends on the dimensions of the robot arm. It should be noted that when the end-effector is attached to the wrist, the work volume exceeds the one given by the robot manufacturer, and this should be taken into account when planning for the safety of the people working near the robot.

The various configurations of robotic arms and their corresponding work volumes are discussed below.

9-2.1 Cartesian Coordinate Robots

The main frame of cartesian coordinate robots consists of three orthogonal linear sliding axes, as shown in Fig. 9-3. The structure can be more rigid if constructed like the work table of a milling machine, but then the ratio between the robot work volume and floor space becomes smaller. In a cartesian robot, the manipulator hardware, the inter-polator, and the control algorithms are similar to those of CNC machine tools. Therefore, the arm resolution and accuracy might also be on the order of magnitude of machine tool resolution, denoted as BLU in previous chapters. An important feature of a cartesian robot is the *equal* and *constant spatial resolution*, namely the resolution is fixed in all axes of motions and throughout the work volume of the robot arm. This is not the case with other coordinate systems as will be shown below.

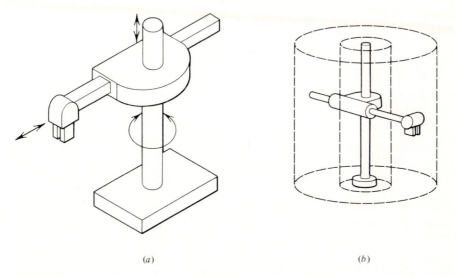

(a)   (b)

Figure 9-4 (a) Cylindrical coordinate manipulator; (b) work volume shape of cylindrical manipulator. (*Courtesy of U.S. Air Force.*)

Figure 9-5 A robot which operates in cylindrical coordinates. The robot can also move in parallel to a moving line. (*Courtesy of Prab.*)

9-2.2 Cylindrical Coordinate Robots

The main frame of cylindrical coordinate robots consists of a horizontal arm mounted on a vertical column, which, in turn, is mounted on a rotary base, as shown in Fig. 9-4a. The horizontal arm moves in and out, the carriage moves up and down on the column, and these two units rotate as a unit on the base. Thus the working volume is the annular space of a cylinder, as shown in Fig. 9-4b. A commercial robot with a cylindrical coordinate motion is shown in Fig. 9-5. This robot, however, can also move parallel to a moving line in order to reach all points within a cube.

The resolution of the cylindrical robot is not constant and depends on the distance ℓ between the column and the wrist along the horizontal arm. If the resolution unit of the rotary base is α radians, then the resolution at the arm end is $\alpha\ell$.

> **Example 9-1** The position-measuring device of the rotary axis of a cylindrical robot is an incremental digital encoder which emits 2000 pulses per revolution and is mounted directly on the shaft. If the maximum length of the horizontal arm is 1 m, what is the worst-case resolution at the arm end?

SOLUTION The resolution at the base is $360/2000 = 0.18°$. The resolution at the arm end is $1000 \times 0.18 \times \pi/180 = 3.14$ mm.

The result of Example 9-1 demonstrates that the arm resolution around the base might be by two orders of magnitude larger than that regarded as the state of the art in machine tools (0.01 mm). This is one of the drawbacks of cylindrical robots as compared to those with cartesian frames. Cylindrical geometry robots do offer the advantage of higher speed at the end of the arm provided by the rotary axis. But this speed is limited in many robots because of the time-varying moment of inertia of the robot arm.

In robots which contain a rotary base, a good dynamic performance is usually difficult to achieve. The moment of inertia reflected at the base drive depends not only on the weight of the object being carried but also upon the distance between the base shaft and the manipulated object. This distance is a function of the instantaneous position of the gripper and the robot joints during the motion. As a result, the effective moment of inertia at the base drive generally varies with time or position. This is regarded as one of the main drawbacks of robots containing revolute joints.

9-2.3 Spherical Coordinate Robots

The kinematic configuration of spherical coordinate robot arms is similar to the turret of a tank. They consist of a rotary base, an elevated pivot, and a telescoping arm which moves in and out as shown in Fig. 9-6a. The magnitude of rotation is usually measured by incremental encoders mounted on the rotary axes. The work envelope is a thick spherical shell as shown in Fig. 9-6b.

The disadvantage of spherical robots compared with their cartesian counterparts is that there are two axes having low resolution which varies with the arm length.

Example 9-2 Find the worst spatial resolution of a spherical robot with 500-mm arm length. The robot is equipped with 3 encoders emitting 1000 pulses per revolution. The linear axis is actuated with the aid of a 20-mm-pitch leadscrew, and its encoder is mounted on the leadscrew.

SOLUTION The linear axis resolution is $20/1000 = 0.02$ mm. The rotary axis resolution is

$$500 \times \frac{360}{1000} \times \frac{\pi}{180} = 3.14 \text{ mm}$$

The spatial resolution is $3.14 \times 3.14 \times 0.02$ mm.

The last example demonstrates the large difference in the obtained resolution between linear and rotary axes. Motions with rotary axes, however, are much faster than those along linear axes. The main advantage of spherical robots over the cartesian and cylindrical ones is a better mechanical flexibility: the pivot axis in the vertical plane permits convenient access to the base or under-the-base level.

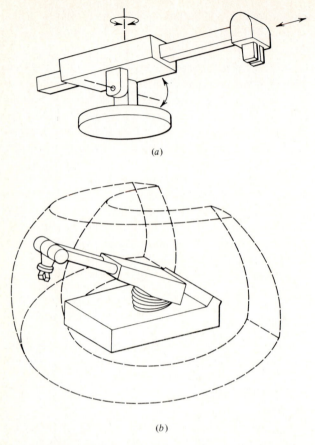

(a)

(b)

Figure 9-6 (*a*) Spherical coordinate manipulator; (*b*) work volume of spherical manipulator. (*Courtesy of U.S. Air Force.*)

9-2.4 Articulated Robots

The articulated, or jointed-arm, robot consists of three rigid members connected by two rotary joints and mounted on a rotary base as shown in Fig. 9-7. It closely resembles

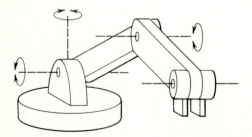

Figure 9-7 Articulated manipulator. (*Courtesy of U.S. Air Force.*)

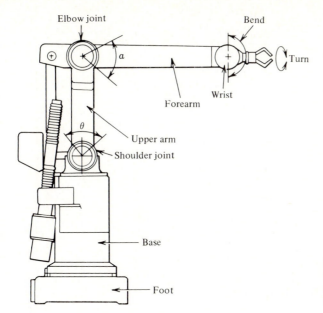

Elbow joint   Bend

a Turn

Wrist

Forearm

θ

Upper arm

Shoulder joint

Base

Foot

Figure 9-8 An articulated robot arm driven through kinematical subchain by translational driving systems. (*Courtesy of ASEA.*)

the human arm. The gripper is analogous to the hand, which attaches to the forearm via the wrist as shown in Fig. 9-8. The elbow joint connects the forearm and the upper arm, and the shoulder joint connects the upper arm to the base. Sometimes a rotary motion in the horizontal plane is also provided at the shoulder joint.

Since the articulated robot has three rotary axes, it has relatively low resolution which depends entirely on the arm length. This is illustrated in Fig. 9-9, which summarizes the spatial resolutions of the four robot configurations. Its accuracy is also the poorest since the articulated structure accumulates the joint errors at the end of the arm. On the other hand it can move at high speeds and has excellent mechanical flexibility, which has made it the most popular medium-size robot.

9-2.5 The Wrist

The end-effector is connected to the main frame of the robot through the wrist. The wrist includes three rotary axes denoted by roll, pitch, and yaw as shown in Fig. 9-1. The roll (or twist) is a rotation in a plane perpendicular to the end of the arm, pitch (or bend) is a rotation in a vertical plane, and yaw is a rotation in horizontal plane through the arm. However, there are applications which require only two axes of motion in the wrist. For example, since the welding gun is a symmetrical tool, most arc welding tasks require a wrist with only two degrees of freedom.

In order to reduce weight at the wrist, the wrist drives are sometimes located at the base and the motion is transferred with chains or rigid links as shown in Fig. 9-10. Reduction of weight at the wrist increases the maximum allowable load and reduces the moment of inertia which improves the dynamic performance of the robot arm.

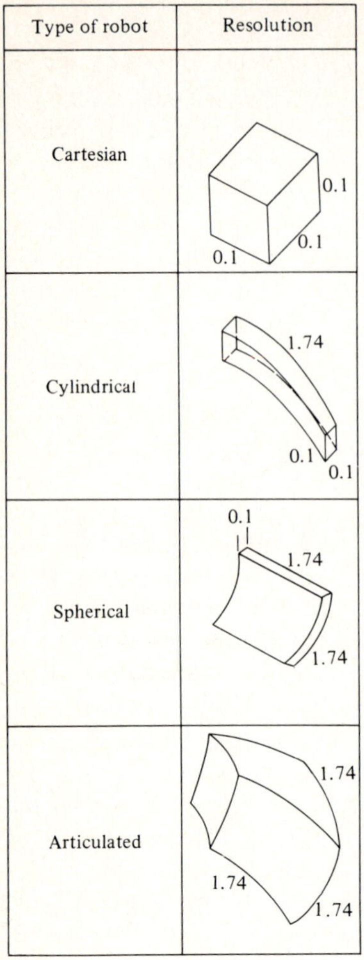

Type of robot	Resolution
Cartesian	0.1 0.1 0.1
Cylindrical	1.74 0.1 0.1
Spherical	0.1 1.74 1.74
Articulated	1.74 1.74 1.74

Figure 9-9 Comparison of spatial resolution in various configurations of a robotic arm. Linear axes resolution, BRU = 0.1 mm; rotary axes resolution, BRU = 0.1°; arm length, 1 m.

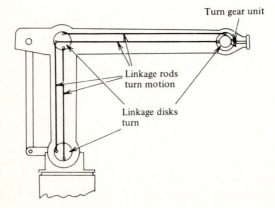

Turn gear unit

Linkage rods
turn motion

Linkage disks
turn

Figure 9-10 Schematic diagram of wrist transmission. (*Courtesy of ASEA.*)

9-3 THE CONTROL AND DRIVES

The control hardware of servo-controlled industrial robots is similar to those of CNC machine tools. Each axis is separately driven, and the required path motion is coordinated by the computer. The computer algorithms, however, are more comprehensive in robotics since they include simultaneous control of six axes of motion, rather than two or three linear axes, as in machine tools. The control program of most noncartesian robots contains also a coordinate transformation routine, as shown in Fig. 9-11 for the articulated-type robot. The input to this routine is the required velocity (V) and the axial positions of the wrist, in the X, Y, and Z directions. The axial positions are transformed to corresponding required angular positions and velocities of the base, shoulder, and elbow joints. These angular values are used as the reference signals to the control loops of the robot arm, as shown for one axis in Fig. 9-11. Similar loops exist for each joint of the robot arm. It should be mentioned that the transformation from the linear to the angular positions is not unique, and an additional algorithm should be incorporated in order to determine the optimal solution. Note that the path between two successive X, Y, and Z values specified, is not predictable. To generate a desired continuous trajectory, one must specify many points, at small intervals, along the trajectory.

Similar to machine tools (see Chap. 1), two basic types of robots are available: point-to-point (e.g., spot welding robots) and contouring, or continuous-path (e.g., spray painting or arc welding robots). Unlike those of machine tools, the controllers of many continuous-path robots do not contain interpolators. The non-cartesian coordinate

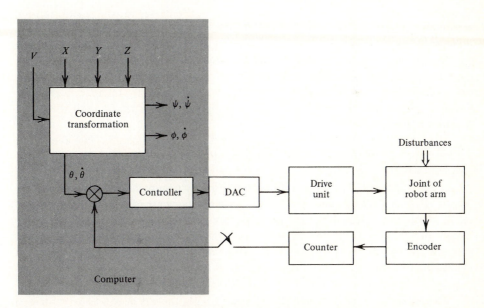

Figure 9-11 Control diagram of an articulated robot (Sampled-Data Method).

system associated with most robots and the presence of three rotary axes at the wrist make implementation of simple interpolation techniques difficult in robotics. Instead, many continuous-path robots use memory devices to store many spatial positions that describe the required path, and these positions are played back during the operation.

9-3.1 Control Loops

In point-to-point robots each axis of motion is usually driven by a dc servomotor or a hydraulic actuator. Each axis includes a position feedback device, where the incremental encoder is the most popular one. The encoder delivers voltage pulses, each corresponds to an axial motion of one basic resolution-unit (BRU). The required position of each axis is expressed in the computer by integers in BRUs. The computer sends equal velocity command signals to all axes, and therefore the axes are always moving at the same volocity. As a consequence, the path of the end-effector while traveling from one point to the next is not controlled. Only the final position at the required point is controlled with the aid of position counters. Each axis is equipped with a position counter which counts the corresponding encoder pulses and stops the axis motion when the required point is reached.

In continuous-path robots, each axis of motion is actuated separately, usually in a closed-loop control. For each motion of the arm, reference signals proportional to the required axial position and velocity are simultaneously sent by the control computer to all axes. Each axis is equipped with a feedback device, usually an incremental encoder, which provides both velocity and position feedback. With the encoder, the number of emitted pulses is proportional to the incremental position of the axis, and the frequency of these pulses is proportional to axial velocity. The reference and the feedback velocity signals are compared; their difference is integrated (to generate the axial position error), amplified, and sent as the command signal to the drive element. At the same time the number of feedback pulses, which represents the actual position, is also counted and compared to the required position. When the required and actual positions are equal, the corresponding axial reference signal is blocked and consequently the motion of the axis stopped at the desired position. The controller then sends new reference signals for the next motion of the manipulator, and the operation is repeated until the entire program has been executed.

Most small- to medium-sized robots utilize dc servomotor drives. Two alternative approaches exist to the control of the motion of a robot arm driven by dc motors [11]. The approach used by several U. S. robot manufacturers and researchers at U. S. universities† is to control the torque of the robot arm by manipulating the motor current. Another approach, commonly used by European and Japanese robot manufacturers and in NC machine tools, is to control the motor rotational speed by manipulation of the motor voltage. The first approach, based on manipulation of current, treats the torque produced by the motor as an input to the robot arm. The second approach, based on manipulatin of voltage, treats the robot arm as a distrubance or load acting on the

---

† Information based on personal communications of the author with V. Scheinman, T. Binford (Stanford University), S. Dubowsky (UCLA), and D. Tesar (University of Florida).

motor's shaft. This basic distinction is not merely a philosophical one and has important practical consequences for the final control system design.

A straightforward approach to the control of robot arm motion is to apply at each joint the necessary torque to overcome friction and dynamic torques due to the moment of inertia. Torque control, based on manipulation of dc motor current, utilizes a current amplifier in the motor's drive unit. The problem with this type of system is the need to have an accurate estimate of the changing moment of inertia at each joint of the robot arm in order to obtain the desired velocity and trajectory. If the actual value of the inertia is smaller than expected, then the torque applied is larger than required. This torque is translated to higher acceleration and consequently higher velocity. This can have disastrous consequences; for example, a part can be struck and broken since the velocity is not zero as desired at the target position. On the other hand, if the inertia is larger than expected, there is a loss of time, since the arm decelerates along distance before the target point and "creeps" toward it very slowly.

An important advantage of the torque control approach is that we can maintain a desired torque or force. This is useful in some robotics applications, for example, press fitting or screwing operations during assembly. Another advantage is that when the robot arm encounters resistance (e.g., strikes a rigid obstacle), it maintains a constant torque and does not try to draw additional power from the electrical source.

The alternative approach is to control the velocity of the robot arm by manipulation of the dc motor voltage, utilizing a voltage amplifier in the motor's drive unit. A similar approach is also usually used in hydraulically driven robots. The main advantage of this approach is that variations in the moment of inertia affect only the time constant of the response but do not result in any disasterous consequences and do not affect the time required to reach the target position. The arm always approaches the target smoothly with a very small speed. The problem with this approach is that the torque is not controlled, and the motor will draw from the voltage amplifier whatever current is required to overcome the disturbance torque. This can lead to burning the amplifier's fuse, when the robot arm encounters a rigid obstacle. Another disadvantage is that this system is not suitable for certain assembly tasks, such as press fitting and screwing, which require a consant torque or force.

The selected control approach should be dependent on the application and the environment in which the robot arm operates. When the arm is free to move along some coordinate (e.g., spray painting robots), the specification of velocity is appropriate. When the robot's end-effector might be in contact with another object in such a way as to prevent motion along a coordinate, then the specification of torque is appropriate.

The computer output in velocity-controlled robots can be transmitted either as a sequence of reference pulses or as a binary word (in a sampled-data system). With the first technique, the computer produces a sequence of reference pulses for each axis of motion, each pulse generating a motion of one BRU of axis travel. The number of pulses represents position and the pulse frequency is proportional to the axis velocity. These pulses can actuate a stepping motor in an open-loop system or be fed as a reference to a closed-loop system, as shown in Fig. 7-3. The reference-pulse technique can be used only when the velocity control approach is applied to the robot arm. Design considerations of these control systems were discussed in Chaps. 6 and 7.

Sampled-data systems can be applied to either the velocity or torque control approach. With the sampled-data technique, the control loop is closed through the computer itself as shown in Fig. 9-11. The typical feedback device is an incremental encoder interfaced with the computer through a counter which is incremented by the pulses received from the encoder. The computer samples the contents of the counter at fixed time intervals. The control program compares a reference word with the accumulated contents of the counter to determine the position error. When the velocity control approach is applied, this error signal is fed at fixed time intervals to a DAC, which, in turn, supplies a voltage proportional to the required axis velocity. Design considerations of this sampled-data system were discussed in Chap. 7. When the torque control approach is applied, the error signal is translated in the computer to a corresponding torque signal, and the latter is used to drive the motor.

9-3.2 Drives for Robots

The drives of servo-controlled robots are either stepping motors, dc motors, or hydraulic actuators. Stepping motors are limited in resolution and power and thus are suitable only for small robots. They also tend to be noisy and, therefore, are seldom used in practice. Hydraulic actuators or motors are well suited for large robots where power requirements are high. They can deliver large power while being relatively small in size. This large power/weight ratio is of extreme importance in robotics, where extra weight tends to deteriorate dynamic behavior. Another advantage of hydraulic actuators is that the piston itself can be used as the moving element of linear axes, thus saving the weight and cost of gears, leadscrews, or other transfer mechanism.

The cost of hydraulic actuators is not proportional to the power delivered, and thus they are expensive for small- to medium-sized robots. They also present some problems in terms of maintenance and oil leakage. DC motors are ideally suited for driving small- to medium-sized robots. They can be designed to meet a wide range of power requirements, are relatively inexpensive, and reliable.

9-3.3 Dynamic Performance

The maximum traveling velocity of the manipulator, the amount of overshoot, and the settling time at the target point are the dominant dynamic parameters in robotics. The traveling velocity and settling time provide the overall speed of operation, and the amount of overshoot can change the shape of the generated path (see Fig. 6-19) or can cause disastrous collisions in assembly when the tool collides with an obstruction.

In robots which contain a rotary base, a good dynamic performance is usually difficult to achieve. The effective moment of inertia at the base depends not only on the weight of the object being carried, but also upon the distance between the base shaft and the object. This distance is a function of the instantaneous position of the end-effector and the other joints during the motion. As a result, the moment of inertia reflected at the base is generally time varying. A similar problem exists with other rotary joints. As a result, the open-loop gains of rotary axes cannot be adjusted to obtain optimal dynamic performance. Each loop gain in the controller must be tuned for the maximum permissible inertia load in order to avoid overshoots over the target point. This tuning degrades the performance when smaller moments of inertia loads are present.

The tuning problem is further explained for a dc servomotor drive. A satisfactory performance is obtained with a reference-pulse system (the velocity control approach) when the open-loop gain K is adjusted according to Eq. (7-11):

$$K = \frac{1}{2\tau} \tag{9-1}$$

where τ is the dominant time constant of the drive unit and is derived from Eqs. (4-11) and (6-3):

$$\tau = \frac{\alpha J R}{K_t K_v} \tag{9-2}$$

where K_t and K_v are the motor torque and voltage constants, respectively, R is the rotor resistance, and J is the moment of inertia given by Eq. (4-15):

$$J = J_r + K_g^2 J_\ell \tag{9-3}$$

where J_r is the moment of inertia of the motor rotor, J_ℓ is the moment of inertia of the load, and K_g is the gear ratio between the load and the motor shaft. If J_ℓ is time-varying, then τ becomes time-varying as well and the condition given by Eq. (9-1) cannot be satisfied. The damping factor ζ of the loop is given by Eq. (7-9):

$$\zeta = \frac{1}{2\sqrt{K\tau}} \tag{9-4}$$

In order to avoid large overshoots, the condition $\zeta > 0.6$ should be maintained. Therefore, the gain K is tuned for the maximum τ, which depends on the maximum expected moment of inertia. This causes, however, large damping factors for small moments of inertia and results in a sluggish transient behavior, which decreases the overall operating speed and increases the errors in continuous-path operations.

9-4 PROGRAMMING

The combination of six axes of motion and a noncartesian coordinate system in which most robots operate makes the programming of a robotic system much more difficult than the part programming in machine tools. At least three programming methods are used in robotics:

Manual teaching
Lead through teaching
Programming languages

These methods are discussed below.

Figure 9-12 Manual programming of a robotic arm. (*Courtesy of ASEA.*)

9-4.1 Manual Teaching

This type of programming is the simplest and the most frequently used in point-to-point robotics systems. Teaching is done by moving each axis of the robot manually, until the combination of all axial motions yields the desired position of the robot. The commands of these axial motions are given by the operator, who uses a series of push buttons on a control box as shown in Fig. 9-12. When the desired position is reached, the operator stores the coordinates of the point in the computer memory. This process is repeated for each required point. In replaying the stored points, each axis moves at its maximum velocity until it reaches the required new coordinate. Consequently, some axes will reach their required coordinate before others, as was illustrated in Fig. 1-5, and the path of the robot end-effector between the points is unpredictable. Therefore this method is useless in continuous-path robots, unless the stored points are fractions of an inch apart.

9-4.2 Lead Through Teaching

One might think that the simplest method to program continuous-path robots is perhaps by physically grasping the robot end-effector and leading it through the desired path at the required speed, while simultaneously recording the continuous position of each axis. There are robots in which this method can be applied by shutting down the power

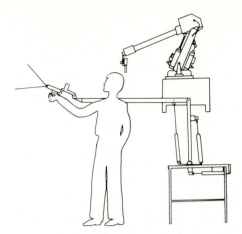

Figure 9-13 Simulator aid in programming.

of the system, e.g., eliminating the oil pressure in hydraulic robots. However, because of the transmission elements (such as gears and leadscrews) in the robotic system, it might be impossible to generate a motion of the robot joints by pulling its end-effector. A solution might be the construction of another identical manipulator, equipped with position feedback devices (usually encoders), but with no drives and transmission elements attached. This device is denoted as the robot simulator and is illustrated in Fig. 9-13. The simulator is manually grasped by the operator through the required path, and at the same time the position of each axis is sampled at a constant frequency and stored in the computer.

While the advantage of this method is the direct programming, it has some major disadvantages:

1. Every operator motion is recorded and played back in the same manner. Therefore unintentional motions will also be played back, unless the system allows the removal of unwanted moves.
2. Since the teaching is performed manually, a high precision in generating paths cannot be achieved.
3. It is impossible to obtain the exact required velocity along the path.
4. A considerable memory size is required to store the data. As an example assume that the position of 6 axes is recorded during 5 min at a rate of 50 samples per second. The amount of required memory words (or bytes) is $6 \times 5 \times 60 \times 50 = 90,000$. For this reason, disk storage devices are usually used with this method.
5. An investment in a simulator is required. Alternatively, if the robot itself can be used in the teaching mode, the robot is tied up during the programming, which consequently reduces its overall efficiency.

The lead through teaching method is widely used in spray painting of auto parts or other products.

9-4.3 Programming Languages

Programming languages in robotics comprises the generation of all data required to move the robot end-effector along a required path in order to perform a specific task. A number of advanced robot control languages are now (1981) in use in research laboratories, such as AL at Stanford University, RPL at SRI International, AUTOPASS and AML at IBM, and PAL at Purdue University. The most advanced commercial language is VAL, which has been designed for use with Unimation, Inc., industrial robots. Programs in VAL are written on the same computer that controls the robot.

To demonstrate the VAL™ language, let us assume that the robot must pick up objects from a chute and place them in successive boxes. One possible sequence of robot activity is as follows:

1. Move to a location above the part in the chute.
2. Move to the part.
3. Close the gripper jaws.
4. Remove the part from the chute.
5. Carry the part to a location above the box.
6. Put the part into the box.
7. Open the gripper jaws.
8. Withdraw from the box.

The corresponding VAL program is as follows:

- EDIT DEMO.1®
- PROGRAM DEMO.1
    1. ?APPRO PART,50®
    2. ?MOVES PART®
    3. ?CLOSEI®
    4. ?DEPARTS 150®
    5. ?APPROS BOX,200®
    6. ?MOVE BOX®
    7. ?OPENI®
    8. ?DEPART 75®

The exact meaning of each line is
1. Move to a location 50 mm above the part in the chute.
2. Move along a straight line to the part.
3. Close the gripper jaws.
4. Withdraw the part 150 mm from the chute along a straight-line path.
5. Move along a straight line to a location 200 mm above the box.
6. Put the part into the box.
7. Open the gripper jaws.
8. Withdraw 75 mm from the box.

When the program is executed, it causes the robot to perform the steps which described the task.

9-5 INTELLIGENT ROBOTS

The servo-controlled robots which have been discussed in previous sections can be programmed to move along a predetermined path. For measuring the displacement they are equipped with position feedback devices, one for each axis of motion. Intelligent robots, which are referred to as the third generation of robots, are using additional sensors, such as TV cameras for vision or force transducers for detecting the disturbing forces acting on the robot end-effector. The addition of sensors allows the robot to deviate from its programmed path and respond to changes in its working environment or, in other words, to make decisions during the operation. In order to utilize the information received by the additional sensors, the robot controller must solve an inverse kinematics problem in real time.

Solution of the direct kinematics problem means calculating the next position and orientation of the robot's end-effector when the six axial position components between the present and next point are given. By contrast, in the inverse kinematics problem the new position and orientation are given; the values of the six axial position components must be calculated. When a vision system is utilized, the distance between the present position of the end-effector (e.g., center of the gripper jaws) and the target point (e.g., object that the robot must pick up) is measured and available in the computer. The present orientation of the end-effector is known; the required orientation at the target point (e.g., pick-up angle of the object) is calculated by a vision algorithm. Based upon

Figure 9-14 A robot interfaced with a vision system identifies parts for assembly. (*Courtesy of Unimation.*)

this information, the computer must provide the six position commands to the six axes of motion, or, in other words, the inverse kinematics problem must be solved in real time. The operating velocity of these robots depends on the computation speed of the inverse kinematics algorithm, and therefore the type of computational method utilized in the control computer is one of the prime features of intelligent robots.

Among the capabilities of intelligent robots is perception and pattern recognition. These robots have the ability to distinguish one shape from another, with the aid of a vision system as shown in Fig. 9-14. The part sensing is executed either by a fixed TV camera interfaced with the robot computer or by a camera built in the robot hand. Future developments include the conversion of spoken languages into operating command, which could lead to manufacturing operations entirely under robot control.

9-6 ECONOMICS

Robots have been manufactured by Unimation, Inc., since 1961, but only in the late 1970s did the demand for robots start to accelerate. The reason for this change to robotics is simple economics. During this period labor costs increased by 250 percent (from \$4 to \$14/h in the automotive industry), while as a result of the declining prices of computers, the cost of a robot has gone up only 40 percent.

The payback period of medium-sized robots, operating in two shifts, is between 1 and 2 years, depending upon the robot cost and type of application. A simple economic analysis [7] assumes that the payback period P is given by

$$P = \frac{R}{L - M} \tag{9-5}$$

where P = payback period in years
R = investment in robot and accessories
L = labor saving per year
M = maintenance and programming cost per year

Example 9-3 The total investment required for a new robot installation is \$70,000, including the price of the robot and tooling. The estimated expense on annual maintenance, robot operating and programming is \$6000 for one shift operation and \$8000 for two shifts. The robot replaces one worker, whose salary (including fringe benefits) is \$24,000. What is the payback period for one-shift and two-shift use?

SOLUTION

One-shift use:

$$P = \frac{70,000}{24,000 - 6000} = 3.88 \text{ years}$$

Two-shift use:

$$P = \frac{70,000}{48,000 - 8000} = 1.75 \text{ years}$$

If the production rate is increased as a result of the robot work, then the payback period is even smaller. Another aspect in cost analysis is the material saving due to the precise operation of the robot compared to human operation. As an example, the saving in paint cost achieved by less overspraying via robot operation can be in the range of 10 to 50 percent depending upon the operation [8]. Additional saving from such things as operator's gloves and toilet facilities, which robots do not need, can also be taken into consideration.

More comprehensive cost analysis might take into account other financial factors like robot depreciation, interest on investment, and increased labor cost. On the other hand, factors such as the risk of accidents due to the failure in robot control should be considered as well.

Related to the economic justification of robots, there are safety and environmental issues which must be taken into account. The use of robots to load and unload presses, for example, eliminates the possibility of accidental operator injury. Equally important is that robots can take operators out of the noisy environments, an issue which is receiving increasing attention today.

9-7 APPLICATIONS OF ROBOTS

The first industrial robots were installed to replace people in dangerous operations, such as loading and unloading hot parts from processing furnaces, or in hazardous environments, for example, when workers were subjected to long exposures of toxic materials. Today, however, industrial robots are primarily installed to improve productivity in manufacturing operations.

The main applications of industrial robots are

Handling and loading of parts
Spray painting
Spot and arc welding
Assembly
Machining operation such as deburring and grinding
Inspection

Each of these applications requires a different level of sophistication of the robotic system, as illustrated in Table 9-1. Let us elaborate on some of these applications.

Machining The most popular machining operations performed by robots are drilling and deburring. The aerospace industry, for example, is using point-to-point robots for drilling holes in aircraft wings.

Table 9-1 Application of robots and type of control

Control type	Application
Point-to-Point	Spot welding Material handling Simple assembly tasks Drilling
Continuous-Path	Arc welding Spray painting Assembly Deburring
Synchronized control with conveyor	Syn. spot welding Syn. loading & unloading Syn. spray painting
Intelligent robot (with sensors)	Loading from a conveyor Complicated assembly tasks Quality inspection

Almost always when machining is performed on metal parts, burrs are generated. The removal of these burrs has usually been done by hand, which is a monotonous and expensive operation. By closely resembling the manual method the industrial robot can

Figure 9-15 Deburring of gear box casing at Volvo, Sweden. (*Courtesy of ASEA.*)

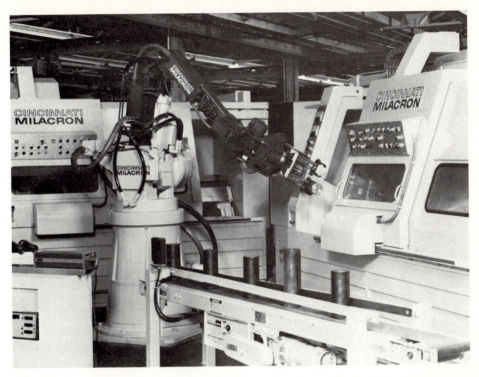

Figure 9-16 A manufacturing cell consists of a T^3 industrial robot and two Cinturn NC turning centers. The robot loads and removes parts from the machines and presents finished parts to a laser gauging station. (*Courtesy of Cincinnati Milacron.*)

solve most deburring problems. An example of how a robot performs deburring is shown in Fig. 9-15. Deburring operations require continuous-path robots.

Handling and loading Industrial robots can be used to transfer heavy parts from one conveyor to another, or to remove finished parts from a machine chuck and place them on a conveyor. In die casting they are used to load inserts into the casting die, and to load and remove castings from the die casting machines. The industrial robot is the main element of the manufacturing cell, in which one robot services from two to four machine tools (see Chap. 10). The robot loads and unloads the machines and transfers the parts between them as shown in Fig. 9-16. Since the loading is assisted by mechanical fixtures, the required positioning accuracy is relatively low.

Spot welding Robots are used to produce spot welds in production lines of the automotive industry and in part assembly operations. The spot-welding robots are usually hydraulically powered, which enables them to carry the welding gun. This gun can weigh up to 70 kg and electric-motor-driven robots cannot usually handle such heavy loads.

The spot welding is a point-to-point operation, in which the traveling path from one point to the next one is not significant. Since the welded points are usually close to one another, the robot's traveling velocity is not a prime requirement; a typical positioning accuracy on the order of 1 mm is required.

Arc welding This is a continuous-path operation, in which both a predetermined path and the velocity along this path should be maintained by the robot. The velocity is determined from welding considerations. The required accuracy along the path depends on the type and size of the welded item.

Spray painting Painting robots are usually hydraulically powered. Likewise in arc welding, painting also requires a continuous-path robot. However, the requirements for a controlled velocity and accuracy along the path are less severe than those with the arc welding robot [8].

Assembly The assembly system is a combination of a robot, a transfer device, and parts feeders. One approach in designing assembly systems is that the assembled object is moving on a straight-line indexing, or rotary indexing, table, and in each station a robot is adding one part to the assembled object. This configuration requires *many simple robots* along the assembly line, each performing a particular operation. The advantage of this system is the small cycle time per product. Another approach applies *one sophisticated robot* surrounded by indexing magazines for the assembly parts. The object is assembled in one location, as shown in Fig. 9-17. During assembly, parts are picked by the robot from the magazines, moved to the assembly location, and assembled.

Sophisticated assembly robots usually possess accurate point-to-point systems with linear continuous-path control. The continuous-path control is needed for the more complicated assembly tasks such as the assembly of a part by sliding it along a grooved path on another part.

In complicated assembly operations both velocity and positioning accuracy are very important. The assembly time is inversely proportional to the robot's velocity. The positioning accuracy requirement is the most severe compared with the other types of robots and in many applications should be on the order of 50 μm.

In order to avoid the demand for high-precision robots, part feeders, and indexing tables, researchers are trying to add sensors to assembly robots. According to this approach the assembly robot should emulate a human operation. The human arm is inaccurate, and humans can assemble an object because they are assisted by vision and tactile senses. Along this line, two approaches are still under study:

1. Using force feedback in an active or passive control scheme to position and orient the parts for assembly accurately. A passive compliance device called *remote center compliance* (RCC) has been developed and demonstrated for this purpose [20]. This device uses a special gripper design where the contact force between misaligned parts causes the RCC device to reposition the part it is holding. The active force feedback control systems also use a special gripper to sense contact

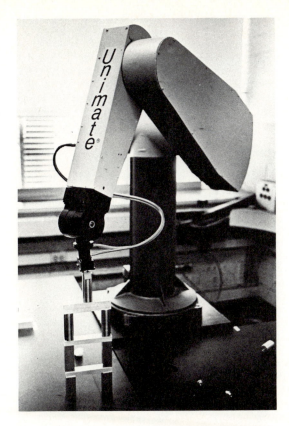

Figure 9-17 A robot performs assembly work in the Center for Robotics and Integrated Manufacturing at the University of Michigan.

forces and torques, and use this information to reposition the gripper actively [4, 18, 20].

2. Adding vision sensors to the robot to enable it to correct its position while picking up and assembling the parts [13, 14]. This is basically an attempt to emulate a human operation in assembly. The main disadvantage of this method is the relatively long time required to process the feedback information, which consequently causes a substantial increase in the assembly time.

Further development in vision systems and force and tactile sensors promise a potential automation of batch assembly operations. However, it is difficult to predict whether these intelligent systems will replace the servo-controlled robots which can perform assembly tasks with the aid of indexing tables and other mechanical fixtures.

BIBLIOGRAPHY

1. Abraham, R. G., J. F. Beres, and N. Yaroshuk: Requirements Analysis and Justification of Intelligent Robots, *Proc. 5th Intern. Sym. Ind. Robots,* pp. 89–111, September, 1975.

2. Abt, H. D.: Flexible Automation by Industrial Robots with CNC, *Ind. Robot,* vol. 3 no. 2, pp. 55–58, June, 1976.
3. Boothroyd, G.: "Handbook on Feeding and Orienting Parts," University of Massachusetts, 1975.
4. Dubowsky, S., and D. T. DesForges: The Application of Model-Referenced Adaptive Control to Robotic Manipulators, *Trans. ASME, J. Dyn. Meas. Contr.,* vol. 101, pp. 193–200, September, 1979.
5. Duffy, J.: "Analysis of Mechanisms and Robot Manipulators," John Wiley and Sons, New York, 1980.
6. Engelberger, J. F.: "Robotics in Practice," AMA/COM, A division of American Management Associations, New York, 1980.
7. ——: Robots Make Economic and Social Sense, *Atlanta Econ. Rev.,* 1977; published in "Industrial Robots," vol. 1, SME, Dearborn, Mich., pp. 35–38, 1979.
8. Fender, N.: Robots in Paint Finishing, *Intern. Eng. Conf.,* 1978; published in "Industrial Robots," vol. 2, SME, Dearborn, Mich., pp. 161–168, 1979.
9. Ford, B.: Industrial Applications for Electrically Driven Robots, *Intern. Eng. Conf.,* 1978; published in "Industrial Robots," vol. 1, SME, Dearborn, Mich., pp. 235–244, 1979.
10. Koren, Y.: "Robotics for Engineers," to be published by McGraw-Hill Book Co., New York, 1985.
11. ——, and A. G. Ulsoy: Control of Servo-Motor Driven Robots, *Proc. Robots VI Conf.,* Detroit, Mich., March, 1982; also published in *ASME, Paper No. 81-WA/DSC-16.*
12. Paul, R.: "Robot Manipulators: Mathematics, Programming, and Control," The MIT Press, Cambridge, 2nd printing, 1982.
13. Rosen, C. A., and D. Nitzan: Exploratory Research in Advanced Automation, NSF Grant G1-38100X1, SRI International *Project reports,* December, 1974, June, 1975, and January, 1976.
14. ——, and D. Nitzan: Machine Intelligence Research Applied to Industrial Automation, SRI International *Project report,* November, 1976.
15. Spur, G., B. H. Auer, and W. Weissel: Handling Automation for Two Milling Machine Tools, *Proc. 5th Intern. Sym. Ind. Robots,* Chicago, Ill., pp. 271–284, September, 1975.
16. Taguchi, N., T. Ishibara, and T. Suzuki: Vertically Mounted Robot for Spot Welding, *Proc. 4th Intern. Sym. Ind. Robots,* pp. 395–404, November, 1974.
17. Tanner, W. R.: Basics of Robotics, Robots II Conf., 1977; published in "Industrial Robots," vol. 1, SME, Dearborn, Mich. pp. 3–12, 1979.
18. Van Brussel, H., and J. Simons: The Adaptable Compliance Concept and its Use for Automatic Assembly by Active Force Feedback Accommodations, *Proc. 9th Intern. Sym. Ind. Robots,* pp. 167–181, March, 1979.
19. Weichbrodt, B., and L. Beckman: Some Special Applications for ASEA Robots—Deburring of Metal Parts in Production, Robots II Conf., 1977, published in "Industrial Robots," vol. 2, SME, Dearborn, Mich. pp. 161–168, 1979.
20. Whitney, D. E.: Force Feedback Control of Manipulator Fine Motions, *Trans. ASME, J. Dyn. Sys. Meas. Contr.,* vol. 99, pp. 91–97, June, 1977.
21. ——, and J. L. Nevins: What Is the Remote Center Compliance (RCC) and What Can It Do?, *Proc. 9th Intern. Sym. Ind. Robots,* pp. 135–152, March, 1979.
22. Wright Aeronautical Laboratories, U.S. Air Force, "ICAM, Robotics Application Guide," Technical Report AFWAL-TR-80-4042, vol. II, April, 1980.

PROBLEMS

9-1 How would a cylindrical robot and a cartesian robot be called by the symbolic notation?

9-2 A robot with a rotary base is located at point O and has to transfer an object from point A to point B, where $OA = OB = AB = 500$ mm. The base drive is a dc motor which rotates at 540 rpm and is connected to the base shaft through a reduced gear ratio of 3:1. An incremental encoder which emits 500 pulses per revolution is attached to the motor shaft. Find the traveling time from A to B and the resolution at point B.

9-3 The transfer of the object in Prob. 9-2 is executed with a cartesian coordinate robot, equipped with a 20-mm-pitch leadscrew, on which the motor and the encoder of Prob. 9-2 are mounted. Find the longitudinal resolution and the traveling time from point A to point B.

9-4 The time constant τ of a robotic arm can change in the range of 20 ms through 42 ms. The design requirement is a damping factor $\zeta \geq 0.65$.

(a) What is the recommended open-loop gain?

(b) What is the damping factor when the smallest τ exists?

(c) With the aid of Fig. 6-11, estimate the ratio between the times required to achieve 90 percent of the steady-state speed with the two extreme time constants.

9-5 Five industrial robots are being considered to replace 12 workers in a production line. Each robot will cost $60,000, including tooling. Annual maintenance for each robot is estimated as $4000. Programming cost for the whole system is $10,000. What is the payback period, based on a one-shift operation, if the workers are paid an annual salary of $20,000, including fringe benefits?

COMPUTER-INTEGRATED
MANUFACTURING SYSTEMS

A quiet revolution is going on in the manufacturing world that is changing the look of the factory. Computers are controlling and monitoring the processes and doing it far more efficiently than human operators. The high degree of automation that until recently was reserved for mass production only, is applied now, with the aid of computers, also to small batches. This requires a change from hard automation in the production line to a flexible manufacturing system which can be more readily rearranged to handle new market requirements.

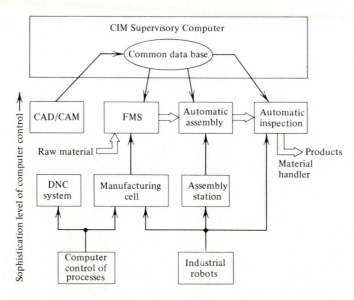

Figure 10-1 Hierarchical structure of computer control of manufacturing systems.

Flexible manufacturing systems combined with automatic assembly and product inspection on one hand, and CAD/CAM systems on the other, are the basic components of the factory of the future. The supervision of this factory, which is predicted to take place by the end of the twentieth century, will be performed by a computer-integrated manufacturing system in which the production flow, from the conceptual design through the finished product, will be entirely under computer control and management.

10-1 HIERARCHICAL COMPUTER CONTROL

The availability of computers in the manufacturing plant has brought a hierarchical structure of computer control to the factory, as illustrated in Fig. 10-1. The lowest level of this structure contains stand-alone computer control systems of manufacturing processes and industrial robots. The computer control of processes includes all types of CNC machine tools, welders, electrochemical machining (ECM), electrical discharge machining (EDM), and a high-power laser as well as the adaptive control of these processes.

When a battery of NC or CNC machine tools is placed under the control of a single computer, the result is a system known as *direct numerical control (DNC)*. The DNC computer is also used for management and inventory control. This system obviously enhances the utilization of computers.

The operation of several CNC machines and a single robot can be incorporated into a system which produces a specific part or several parts with similar geometry. This structure, which substitutes for the DNC system, is denoted as the *manufacturing cell*. The computer of the cell is interfaced with the computers of the robot and the CNC machines. It receives "completion of job" signals from the machines and issues instructions to the robot to load and unload the machines and change their tools. The software includes strategies permitting the handling of machine breakdown, tool breakage, and other special situations.

The operation of many manufacturing cells can be coordinated by a central computer via the aid of a material-handling system. This the highest hierarchical level in computer control of a manufacturing plant and is denoted as a *flexible manufacturing system (FMS)*. The FMS accepts incoming workpieces and processes them, under computer control, into finished parts.

The parts produced by the FMS must be assembled together into a final product. In the factory of the future the parts will be routed on a transfer system to assembly stations. In each station, a robot will assemble parts either into a subassembly or (for simple units) into the final product. The subassemblies will be further assembled by robots located in other stations. The final product will be tested by an automatic inspection system.

Another type of computer utilization is the computer-aided design/computer-aided manufacturing (CAD/CAM) system. The FMS uses CAD/CAM systems to integrate the design and manufacturing of parts in order to minimize the production cycle in the factory. A criterion for design would be the product's suitability for robot assembly.

In this case the CAD/CAM computer must have data associated with the assembly plant.

At the highest hierarchical level there will be *computer-integrated manufacturing (CIM) systems*. Such systems call for the coordinated participation of computers in all phases of a manufacturing enterprise: the design of the product, the planning of its manufacture, the automatic production of parts, automatic assembly, automatic testing, and of course the computer-controlled flow of materials and parts through the plant. All these phases must be integrated into one computer network, supervised by the CIM central computer which monitors the interrelated tasks and controls each of them based on an overall management strategy.

The development of CIM systems is possible because of the improving capability of computer technology, coupled with the pressure for higher productivity, a combination that is resulting in a new era in manufacturing. While CNC manufacturing systems and present robots are replacing the human's *power and skill*, CIM systems will be replacing the human's *intelligence* and using it with incomparably higher efficiency.

The increase in productivity associated with CIM systems will not come due to the speeding up of the machining operations. The breakthrough in productivity is due to the concept of a computer managing an integrated manufacturing system. The CIM computer will be able to make decisions and adapt the production flow to variations in the environment. For example, if a specific product fails in the final inspection station, another similar product must be automatically manufactured in order to meet the requested quantity of output products. The CIM computer will issue an instruction to the manufacturing system to produce the additional corresponding parts required for the assembly of the product. The major increase in productivity associated with CIM systems will be achieved by minimizing the direct labor employed in the plant. In addition to the direct savings in labor costs, one can expect substantial savings from reduced inventories. Far fewer parts, whether finished or in process, will be waiting, either for the next operation or for assembly. The anticipated result is an overall cost reduction by a factor of from 5 to 10 [4]; in other words, a cost reduction of from 80 to 90 percent.

The continuation of this chapter provides the concept of DNC systems, but it is mainly devoted to the description of some of the components of a CIM system: the manufacturing cell (which replaces the DNC), the FMS, CAD/CAM, and management systems, which together draw the outline of the factory of the future.

10-2 DNC SYSTEMS

The term DNC refers to a system of several machine tools directly controlled by a central computer. The first DNC systems were designed in the late 1960s with the idea of eliminating substantial hardware from the individual controller of each machine tool and compensating for the eliminated features by providing a sophisticated central computer. These DNC systems operated in a time-shared mode, with a supervisory program in the central computer linking the machine tools' controllers and establishing any necessary priority.

This solution, however, proved to be economically unjustified, since there were many problems that initially had not been considered. For example, the lengthy cable runs needed for the transmission of the machine commands from a remotely located central computer to the individual machine control units of the DNC system strongly affect the initial investment in the system. System maintenance is another problem, inasmuch as the DNC system, including the central computer as well as the local controllers, is a special-purpose system. This means that software logic, timing problems, and interfacing vary from one system to another. One of the major disadvantages of early DNC system design was that it was difficult to add or remove a machine tool because such changes meant modifications of the machine's controller. These considerations were of great significance to most potential users, and therefore such systems were not generally accepted.

In the late 1970s, DNC manufacturers adopted another concept in DNC system design. The new DNC system circumvents the last problem, as well as the economic and software ones, in that it leaves the conventional NC or CNC system as it is, except

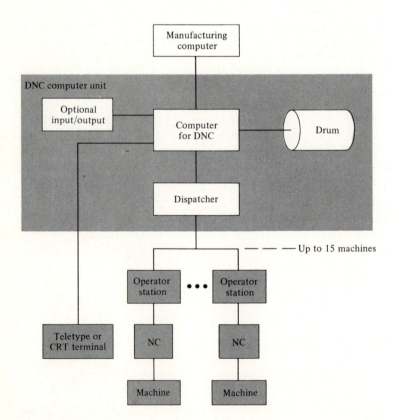

Figure 10-2 General Electric's CommanDir DNC system.

for the tape reader. In this design, the part programs are stored in the central computer, and the only interconnections between the computer and an individual NC controller are those required to simulate the operation of the reader. This approach is known as the *behind-tape-reader interface*. This is the only interface that can be easily adapted to any type of machine tool and that permits any machine tool of the DNC system to be taken out and used as a stand-alone NC machine just by the addition of a tape reader circuit.

General Electric's CommanDir DNC system (see Fig. 10-2), although discontinued, may serve as an example of a DNC design. The system, which was capable of simultaneously operating up to 15 conventional NC machine tools, included a small CPU with a large drum storage and a telecommunication printer. The heart of the processor was a general-purpose minicomputer whose core memory had a cycle time of 1 μs and a capacity that ranged from 16k to 64k bytes. Matching information was supplied to each numerically controlled machine by a data dispatcher linked to the computer. Although the control system bypassed each tape reader, the machine tools could be operated individually.

Since the control of machine tools alone has been economically unsound, DNC manufacturers are using the central computer also as an aid to management. This approach transforms the DNC into a technique that allows machining processes to be fully integrated with other automation systems, with data files, and with management information systems. The aid to management might be considered as the most important advantage of a DNC system.

10-3 THE MANUFACTURING CELL

The manufacturing cell concept was introduced by the Production Engineering Laboratory at the University of Trondheim in Norway [3, 10]. To understand the cellular concept it is necessary to understand certain current economic and social conditions in Norway. Norway is a country that has a small and scattered population which is desirable for national security reasons. The rural areas of Norway suffer from severe unemployment, while in its cities there is a growing shortage of workers willing to work in its manufacturing industry located there. For security and economic reasons, the Norwegian government does not want the rural population to move to the cities. The reasonable solution is to bring miniplants of manufacturing to the rural areas—this was the original cellular concept.

Under this cellular concept, the manufacturing operations are broken down into cells, each at a different plant. Each cell is responsible for the manufacture of a specific part family, namely, parts with similar features, determined by group technology principles. The cells are interconnected by a network of material and finished parts transports.

However, Norwegian wages are among the highest in the world. Therefore, in order for the Norwegian manufacturing industry to be competitive in the world market, its productivity must be high. This brought the second concept in the plan: to provide each miniplant with a high-technology core.

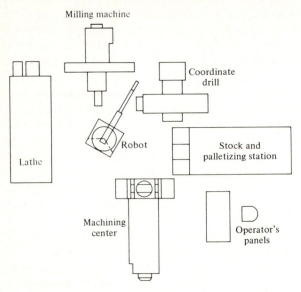

Figure 10-3 The core of a manufacturing cell.

The core consists of a group of several (two to five) CNC machine tools arranged in a circle around a single robot, as shown in Fig. 10-3. The robot does all the part handling and machine loading and unloading in the cell. The supervision and the coordination among the various operations is performed by the cell computer, which performs a similar task to that of the DNC computer. The core runs 24 h continuously but requires workers' participation only during the day. The day shift prepares the computer programming, production planning, scheduling, heat treatment, mounting parts on pallets, etc.

The utilization of the cellular concept does not necessarily require that the manu-facturing cells be located great distances from each other. On the contrary, a higher productivity can be achieved if all the cells are located along a single material transfer system, such as a long conveyor, on which raw workpieces, and semifinished and finished parts are moving. A "ready for workpiece" signal from the control unit of the first machine in a manufacturing cell would instruct the robot to look for the required workpiece on the conveyor. The robot will pick up the workpiece, load it onto the machine, and send a signal to the machine control to begin its operations on the workpiece. While waiting for completion or transfer of the part to the next machine, the robot could perform housekeeping functions such as chip removal, staging of tools in the tool changer, and inspection of tools for breakage or excessive wear; in all of these chores, other interconnect signals could alter the functions performed by the robot depending on the outcome of these tests or the presence of any unusual situations during this housekeeping. If during these functions a machine control detects a malfunction or a tool breakage during a machine operation, the robot must abandon these routine tasks and take some action to either remedy the problem or initiate an emergency procedure for the total system. A "part finished" signal from the last machine tool to the robot would request that the finished part be unloaded and transferred to the outgoing conveyor. The cycle could then be repeated.

Manufacturing cells, or machining cells, are in operation in industry. In Fanuc, Inc., a Japanese producer of CNC controllers and robots, the production is performed in cells, each consisting of two or three machine tools and a rotary indexing table, arranged in a circle around a robot. Each indexing table contains initially raw workpieces. The robot picks up a workpiece, loads it onto the first machine, and later transfers it among the machines for processing. At the end of the process the robot returns the finished part to the same location on the indexing table from which the workpiece has been taken. The table then moves one station and the cycle is repeated. At the end, the indexing table contains only finished parts.

In the United States, Wasino Corp. installed a cell consisting of two CNC lathes and a double-gripper robot [13]. In a typical sequence, the robot picks up a workpiece from a table with one of its two grippers and approaches the spindle of the first machine. There the robot retrieves with its second gripper the partially machined workpiece from the spindle, rotates its wrist 180°, and substitutes the new workpiece from its other gripper. The new workpiece is clamped, the robot withdraws, and the machining begins. The robot then moves to the second machine and substitutes the finished part at the spindle with the partially machined workpiece which was in its gripper. Finished parts are lifted out by the robot and deposited in a hopper.

Another similar manufacturing cell was installed by Olofsson Corp. at Deere and Co. [13]. It consists of two Model 75 CNC vertical precision turning machines, a head center, and a robot. Also in this cell the robot is double-handed and can handle two parts at a time.

The optimization of the production rate in a manufacturing cell requires a different strategy than the one applied to a single machine tool. In a manufacturing cell, only one machine can operate with a cutting speed derived by the conventional minimum production time criterion (see Sec. 8-6), or with an AC system which maximizes the material removal rate (see Secs. 8-3 and 8-5). This machine should be the slowest one in the production process of the part in the cell. One reasonable optimization strategy of a cell could be an equal machining time on each machine tool. With this strategy the product cycle time equals the machining time, plus the relative portion of the tool changing time, plus workpiece loading and unloading times at the spindle by a double-gripper robot.

10-4 FLEXIBLE MANUFACTURING SYSTEMS

The various individual manufacturing systems that were introduced throughout this book can be incorporated into a single large-scale system in which the production of parts is controlled with the aid of a central computer. The advantage of such a production system is its high flexibility in terms of the small effort and short time required to manufacture a new product, and therefore it is denoted as a flexible manufacturing system (FMS).

10-4.1 The FMS Concept

The FMS provides the efficiency of mass production for batch production. The term *batch production* is applied to parts manufactured in lots ranging from several units to more than 50, for which the total annual demand is fewer than, say, 100,000 units. The term *mass production* applies when higher annual production rates are required, and then the use of special-purpose machines can be justified. To machine a single unit with general-purpose machine tools may cost 100 times as much as to manufacture the same part by the most efficient mass production methods [4]. As an example, consider a complex mass-produced part with which almost everyone is familiar: the cylinder block for a typical V-8 automobile engine. Under mass production conditions, where the engine block is conveyed automatically along a transfer line, with the various operations (drilling, tapping, boring, milling, and so on) being executed in sequence at the different stations along the line, the complete machining cost (excluding the raw material) would be on the order of $25. If, however, only a few special cylinder blocks were to be made with machine tools and skilled labor, the machining cost per block could easily rise from $25 to $2500 or more [4].† By making use of FMS technology it should be possible to reduce the cost of producing parts in small and medium quantities.

Existing FMSs in the United States are typically made up of machining centers working in concert with other types of machines, all under the control of a central computer. The workpieces are on pallets which move throughout the system, transferred by towlines (or dragchains) located beneath the floor or by some other mechanism. These FMSs limit handling by the operators and can be more readily reprogrammed to handle new requirements.

Future FMSs will contain many manufacturing cells, each cell consisting of a robot serving several CNC machine tools or other stand-alone systems such as an inspection machine, a welder, an EDM machine, etc. The manufacturing cells will be located along a central material transfer system, such as a conveyor, on which a variety of different workpieces and parts are moving. The production of each part will require processing through a different combination of manufacturing cells. In many cases more than one cell can perform a given processing step. When a specific workpiece approaches the required cell on the conveyor, the corresponding robot will pick it up and load it onto a CNC machine in the cell. After processing in the cell, the robot will return the semifinished or finished part to the conveyor. A semifinished part will move on the conveyor until it approaches a subsequent cell where its processing can be continued. The corresponding robot will pick it up and load it onto a machine tool. This sequence will be repeated along the conveyor, until, at the end of the route, there will be only finished parts moving. Then they could be routed to an automatic inspection station and subsequently unloaded from the FMS. The coordination among the manufacturing cells and the control of the part's flow on the conveyor will be accomplished under the supervision of the central computer.

Advanced FMSs will contain a high-power laser station incorporated in the production line. The laser will be used mainly for heat treatment, sheet metal cutting,

---

† Although the absolute costs have been changed, the cost ratio is still valid.

drilling, and welding. At present, laser treatment of materials with CO_2 lasers in the 5 to 15 kW range are becoming more popular in the industry. The central computer of advanced FMSs will contain a machining data base to provide the recommended cutting parameters to the machine tools in the plant, based upon a selected tool, workpiece material, and upon maximization of the production rate in the entire plant.

The advantages of FMSs include the following:

1. Increased productivity
2. Shorter preparation time for new products
3. Reduction of inventory parts in the plant
4. Saving of labor cost
5. Improved product quality
6. Attracting skilled people to manufacturing (since factory work is not regarded as boring and dirty)
7. Improved operator's safety

Additional economic savings may be from such things as the operators' personal tools, gloves, etc. Other savings are in locker rooms, showers, and cafeteria facilities—all represent valuable plant space, which will not require enlarging if company growth is achieved with flexible automation systems.

10-4.2 Transfer Systems

FMSs and assembly systems utilize material transfer systems, on which the work-pieces, finished parts, or tools are moving between the machining centers, the manufacturing cells, or assembly stations. Various types of transfer systems have been utilized in existing FMSs and a few of them are discussed below.

Towline system. In this system workpieces are attached to pallet fixtures, or platforms, which are carried on carts that are towed by a chain located beneath the floor. The pallet fixture is designed so that it can be conveniently moved and clamped at successive machines in manufacturing cells. The advantage of this method is that the part is accurately located in the pallet, and therefore it is correctly positioned for each machining operation.

Wire-guided carts. Two carts, which are moving along paths determined by wire embedded in the floor, are shown in Fig. 10-4. In the foreground, the cart has picked up a finished palletized workpiece from the machining center in the background and delivers it to an unload station elsewhere in the system. The cart in the background is about to deliver a rack of replacement tools to the receiver station on the machining center in the foreground. Tool interchange from this rack to the machining center's storage chain is automatic and computer-controlled.

Roller conveyor system. In this system a conveyor consisting of rotating rollers runs through the factory. The conveyor can transport palletized workpieces or parts which

Figure 10-4 Material handling by wire-guided carts in a FMS. (*Courtesy of Cincinnati Milacron, Inc.*)

are moving at constant speed between the manufacturing cells. When a workpiece approaches the required cell, it can be picked up by the robot or routed to the cell via a cross-roller conveyor. The rollers can be powered either by a chain drive or by a moving belt which provides the rotation of the rollers by friction.

Belt conveyor system. In this type of transfer system either a steel belt or a chain driven by pulleys is used to transport the parts. This general type of transfer system can operate by three different methods:

1. *Continuous transfer* In this type the workpieces are moving continuously and the processing is either performed during the motion or the cell's robot picks up the workpiece when it approaches the cell. The in-motion processing transfer system is used in automobile assembly lines in which human work and robot operation are combined. Figure 10-5 illustrates the operation of spot welding robots in car assembly, where the welding is performed while the cars are moving on the transfer line. The speed of the transfer line is taken into account by the robot controller when the welding head is moved from one point to the next.

2. *Synchronous transfer* This method is mainly used in automatic assembly lines. The assembly stations are located with the same distance between them, and the parts to be assembled are positioned at equal distances along the conveyor. In each

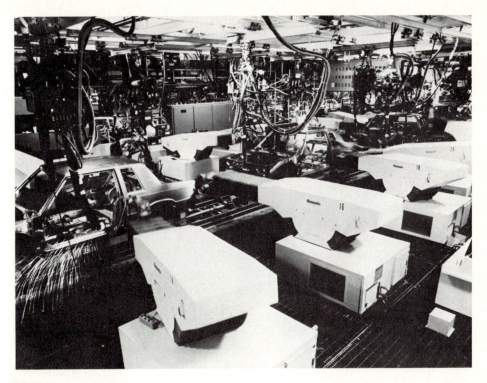

Figure 10-5 Robots weld in an automobile assembly line. (*Courtesy of Unimation, Inc.*)

station a few parts are assembled by a robot or automatic device with fixed motions. The conveyor is of an indexing type; namely, it moves a short distance and stops when the product is in the stations, and subsequently the assembly takes place simultaneously in all stations. This method can be applied where station cycle times are almost equal.

3. *Power-and-free* This method allows each workpiece to move independently to the next manufacturing cell for processing. Usually the method is used for large workpieces and when the manufacturing stations have varying cycle times. The main overhead conveyor loop in Fig. 10-6 is of the power-and-free type. In this case it is used to transfer heavy machining heads between machining stations.

10-4.3 Head-changing FMS

One of the operating FMSs is the Bendix FlexChanger [12], a head-changing system, which is illustrated in Fig. 10-6. It consists of 3 machining stations, 77 machining heads, and 11 probe heads. Parts produced are cast iron transmission cases and reverser cases, each weighing approximately 90 kg.

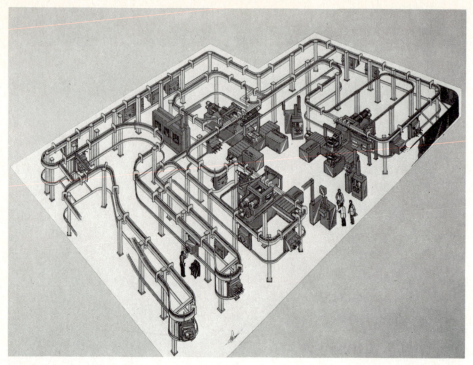

Figure 10-6 Head-changing FMS. (*Courtesy of Bendix, Inc.*)

In the FlexChanger system, the parts and fixtures remain at the machining stations, and the machining heads are transferred to and through the stations as required so that work can be performed in any combination. Fifty heads are used when running transmission cases, and 38 heads are required for the reverse cases.

Operations at each machining station are controlled by a CNC unit. A programmable controller incorporated in the CNC is used to control and interface conveyor loop operation with the machining cycle. Operations performed include drilling, tapping, spotfacing, reaming, rough boring, finish milling, grooving, and probing. Tools in machining heads range from a single-spindle mill to a multiple-spindle head incorporating 23 different tools. The probe heads check for broken tools and include an air blowoff to clean out loose chips and other foreign material.

Heads that are not in use at any of the three machining stations are stored in a main overhead conveyor loop of the power-and-free type. Heads not immediately required for machining are carried in a second overhead storage loop, 6 m off the floor. Each loop can hold up to 55 machining heads for a total of 110. Total length of both loops is about 430 m (1400 ft).

In addition to the two conveyor loops, the head transport system includes a standard roll-type transfer bar at each machining station. As a machining head approaches a station where it is to machine a part, the head is automatically disengaged from the conveyor chain drive mechanism. The transfer bar then simultaneously moves this head into position and the head which has completed its machining operations moves out and back into the conveyor loop.

The incoming head is accurately located and automatically clamped to the feed unit at the machining station. Four locating pads used for positioning incorporate air blowoff to clean chips and other debris from the mounting surface. The air blowoff also serves as a signal to confirm that the head is properly mounted against the locating pads. Then the head is advanced toward the part and the machining begins. After completing its machining operations, the head is backed away from the part and transferred to the conveyor loop.

Changeover from one part to another is done sequentially. As the last part in a run is finished at each machining station, the fixture and CNC program are changed. Machining heads for that part are routed into storage on the conveyor, and heads for the other part are brought on-stream out of storage.

The complete system includes a qualifying machine and two finishing machines— one for each type of part. The qualifier mills several locating surfaces and bores two locating holes. Finish bores and finish mill surfaces are completed in the finishing machines.

10-4.4 Variable Mission Manufacturing System

A FMS denoted as a variable mission manufacturing (VMM) system was developed by Cincinnati Milacron and is in operation in one of its plants.

The VMM system's basic components, as seen in Fig. 10-7, include two CIM-Xchanger 20 HC CNC machining centers being supplied continuously with palletized parts and replacement tools by remotely controlled, wire-guided carts. The carts, following electrical signals from wire embedded in the floor, are moving along a closed-loop layout between and around the two machining centers, as well as into and out of pallet loading and unloading stations. The cart movement and the entire system management is under the control of two computers.

There are seven different pallets moving through the system at all times. Five of these pallets are fixtured to carry different production parts. They move from load stations through the layout to a pallet delivery station at one or the other of the machining centers. The parts then move to the machine's rotary index work module and are clamped automatically in position for machining. Following the machining, each part is moved from the work module to a pallet discharge station where it is picked up by one of the carts and moved on through the system to one of two load-unload stations.

Two of the seven pallets carry a tool rack holding twelve tools. These are delivered intermittently to each of the machining centers, where automatic tool interchange was programmed and tools changed between the rack and the machine's storage chain. These tool changes could be made in a system during an unmanned second or third shift operation. Special features of the system include utilization of a surface sensing probe, which is stored in the tool chain like any other typical tool and brought to the spindle nose on command from the control. The probe can identify true work position and/or other part shape anomalies (such as casting variations). It can feed information back to the computer control of the system to correct for differences or errors. It can detect

Figure 10-7 Variable mission manufacturing system. (*Courtesy of Cincinnati Milacron, Inc.*)

incomplete machining operations caused by broken tools or worn tools and can detect missing surfaces.

The VMM system includes an optional adaptive control feature called *torque-controlled machining*. Microprocessor-based, this feature increases productivity by acting as an adaptive control of feedrate, which depends on torque at the tool and horsepower at the spindle motor. It can sense such conditions as air gaps, workpiece hardness, hole breakthrough, cross and blind holes, and chip clogging.

The wire-guided carts offer significant advantages in flexibility in that it is relatively simple to add, remove, and/or reroute carts at any time. The carts can carry up to 4000 lb. They are battery-powered with sufficient energy storage capability to easily serve an 8-h shift.

The overall VMM control system has been designed in a modular fashion, with individual modules responsible for separate functions. The modules provided are these:

Data distributor   Sends data to and receives data from the machine tool control. The data is normally in the form of NC part programs but can also be tooling data or operator messages.

Data manager   Stores and provides the capability for the user to manage the data in the system. It is also responsible for data transmission to and from the user's remote computer.

Traffic coordinator Controls workpiece movement between work stations within the VMM system.

Work preparation Guides the VMM operator in the activities required prior to entering parts into the system. These activities include work order definition, determination of lot sizes, and the entry of start-stop dates.

Tool manager Stores, monitors, and updates tool data files which contain tool length and diameter compensation, tool life expectancy, etc.

The control of the VMM system is essentially a two-level system. The two machining centers are interfaced with their usual CNC consoles which control the typical machine functions, such as positioning and machining cycles. The CNCs receive their data via computer, however, and not from punched tape. On another level is the overall management of the VMM system. The hardware for this level includes two computers, a video display terminal, two disk drives, and a hard copy console. The computers are a PDP-11/34, which provides overall control of the system, and a PDP-11/35, which provides control over the cart movement. There is also a shop terminal interfaced with the PDP-11/35. This terminal is used primarily to enter commands in a semiautomatic mode. With memory added to this unit, it can be used as a behind-tape-reader interface, serving as hard-wired machine controls to the VMM system.

10-5 CAD/CAM SYSTEMS

Computer-aided design (CAD) means the use of a computer to assist in the design of an individual part or a system, such as an aircraft. The design process usually involves computer graphics.

Computer-aided manufacturing (CAM) means the use of a computer to assist in the manufacture of a part. CAM can be divided into two main classes: (1) On-line applications, namely, the use of the computer to control manufacturing systems in *real time*, such as CNC and AC systems of machine tools. (2) Off-line applications, namely, the use of the computer in production planning and non-real-time assistance in the manufacturing of parts. Examples of off-line CAM are the preparation of part programs on punched tapes by using the APT language (see Chap. 3), or the display of the tool path in machining simulation, as illustrated in Fig. 10-8.

CAD/CAM is a unified software system, in which the CAD portion is interfaced inside the computer with the CAM system. The end result of current CAD/CAM systems is usually a part program in the form of a list or a punched tape. In advanced CAD/CAM systems part programs can be directly fed into the control computers of CNC machines and inspection stations.

10-5.1 Computer-Aided Design

A CAD system is basically a design tool in which the computer is used to analyze various aspects of a designed product. The CAD system supports the design process at

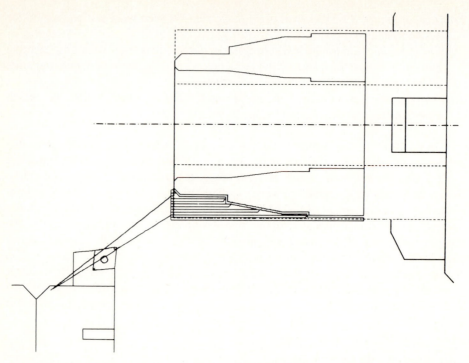

Figure 10-8 A computer display of a turning path on a CAD/CAM system. (*Courtesy of Intergraph Co.*)

all levels—conceptual, preliminary, and final design. The computer can calculate various features of the product, such as its strength, stiffness, and weight. The designer can then test the product in various environmental conditions, such as temperature changes, or under different mechanical stresses.

Although CAD systems do not necessarily involve computer graphics, the display of the designed object on a screen is one of the most valuable features of CAD systems. The picture of the object is usually displayed on the surface of a cathode-ray tube (CRT). Computer graphics enables the designer to study the object by rotating it on the computer screen, separating it into segments, enlarging a specific portion of the object in order to observe it in detail, and studying the motion of mechanisms with the aid of kinematic programs. For example, Fig. 10-9 shows a graphics simulation of a truck running over an antisymmetric bump in the road. The truck and its suspension are simulated in detail, including geometric, inertial, stiffness, bushing, and damping effects. Figure 10-9 displays the truck at time $t = 0$, but other graphs showing the vehicle crossing the bump can be displayed. Such graphics tests can save the expense involved in producing a prototype of the designed object.

Most CAD systems are using interactive graphics systems. Interactive graphics allows the user to interact directly with the computer in order to generate, manipulate, and modify graphic displays. Interactive graphics has become a valuable tool, if not a necessary prerequisite, of CAD systems.

The end products of many CAD systems are drawings generated on a plotter

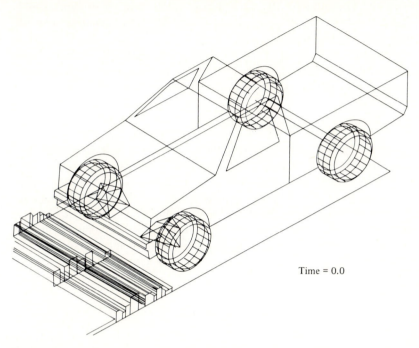

Time = 0.0

Figure 10-9 CAD simulation of a truck running over a bump in the road. (*Courtesy of Mechanical Dynamics, Inc.*)

interfaced with the computer, as illustrated in Fig. 10-9. One of the most difficult problems in CAD drawings is the elimination of hidden lines. The computer produces the drawing as a wire frame diagram. Since the computer defines the object without regard to one's perspective, it will display all the object's surfaces, regardless of whether they are located on the side facing the viewer or on the back, which normally the eye cannot see.

Various methods are used to generate the drawing of the part on the computer screen. One method is to use a geometric modeling approach, in which fundamental shapes and basic elements are used to build the drawing. The lengths and arc radii of the elements can be modified. For example, a cylinder is a basic element; the subtraction of a cylinder with a specific radius and length will create a hole in the displayed part. Each variation, however, maintains the overall geometry of the part.

Other CAD systems use group technology in the design of parts. Group technology is a method of coding and grouping parts on the basis of similarities in function or structure or in the ways they are produced. Application of group technology can enable a company to reduce the number of parts in use and to make the production of parts and their movement in the plant more efficient.

Recently CAD systems are using the finite-element method (FEM) of stress analysis. By this approach the object to be analyzed is represented by a model consisting of small elements, each of which has stress and deflection characteristics. The analysis requires the simultaneous solution of many equations, a task which is performed by the computer. The deflections of the object can be displayed on the computer screen by generating animation of the model.

With any of these methods, or others which are used, the CAD system generates at the design stage a single geometric data base which can be used in all phases of the design and later in the manufacturing, assembly, and inspection processes.

10-5.2 The CAD/CAM Concept

The main concept of CAD/CAM systems is the generating of a common data base which is used for all the design and manufacturing activities (see Fig. 10-1). These include specifications of the product, conceptual design, final design, drafting, manufacturing, and inspection. At each stage of this process, data can be added, modified, used, and distributed over networks of terminals and computers. The single data base provides a substantial reduction in human errors and a significant shortening of the time required from the introduction of a concept of a product to the manufacturing of the final physical product.

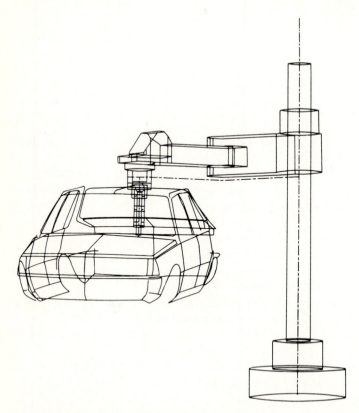

Figure 10-10 Computer determination and display of the trajectories of a robot carrying out spot welding of an automobile body. (*Courtesy of Renault, France.*)

The size and capability of the required computer system depend on the complexity of the product. In the aerospace and aeronautics industry, where a complete aircraft can be designed with a CAD/CAM processor, the system must accommodate new data and changes in data arriving from a variety of users. Therefore these systems must have a strong data management capability. By contrast, if simple products are designed by a company, the required CAD/CAM system would need only one computer terminal. Today the major users of CAD/CAM systems are the aerospace and automotive industries, but the declining prices of these systems enlarge the number of other users.

Advanced CAD/CAM systems include solid geometry modeling capability, in addition to the wire frame model diagrams. The solid description is important when NC verification programs are included in the CAM system, and the changing status of the workpiece during machining simulation should be observed.

A different application of a CAD/CAM system is demonstrated in Fig. 10-10. Fixtures in the operating neighborhood of a robot are displayed on the CRT screen together with the potential movements of the robot. This application allows the designer to predict dangerous interference. Consequently the installment of machines and the design of fixtures can be planned to avoid collisions during the robot operation.

In recent years CAD/CAM technology has improved industry productivity. It is a significant step toward the design of the factory of the future.

10-6 COMPUTER MANAGING SYSTEMS

Studies indicate that in standard manufacturing, a part spends about 5 percent of the time being machined and 95 percent waiting and moving [4,9]. The breakthrough in productivity in the factory of the future will come mainly by cutting down the 95 percent waiting time substantially. This reduction will be achieved by using a computer for managing an integrated manufacturing system.

In managing manufacturing systems the computer functions in both off-line and on-line (or real-time) modes. In the off-line mode the computer is the main tool in resource planning, scheduling, production planning, inventory management, etc. In the on-line mode the computer manages the production flow through the manufacturing and assembly lines.

The realization of a fully computer-integrated manufacturing system is the long-run goal of the industrial society. The strategy being followed to get to this goal is to develop a series of smaller software systems, which can, in the long run, be readily interfaced with each other to build up full systems. An example of an off-line management system for production scheduling and planning is described below.

10-6.1 Materials Management/3000 System

Materials Management/3000 is an interactive material requirements planning and control system designed by Hewlett-Packard. The object of this system is to make it easier to deal with the complexities of operating a manufacturing company by taking into account various aspects of information flow and management. It is primarily designed for manufacturers who build standard products to stock in discrete manufacturing steps (fabricators and assemblers). These companies have a significant investment in inventory, and this system can help them balance their inventory levels with customer demands for timely shipments.

Materials Management/3000 consists of 10 major modules (see Fig. 10-11):

1. *Master production scheduling* (*MPS*) This is a management planning and production scheduling tool. It is used by the master scheduler to generate a production schedule for the plant's marketable products. Input to the module includes current customer orders, forecast customer orders, the current production schedule, and the current level of product inventory. The output of the MPS calculations is called the master production schedule. This schedule contains suggested manufacturing orders including quantities and starting dates. The schedule is then input to the materials requirements planning (MRP) module to plan the manufacture and purchase of the required component parts.

2. *Rough-cut resource planning (RRP)* This is a management tool used by the master scheduler to help produce a realistic master schedule by comparing the resources needed to implement the master schedule with the available critical resources.

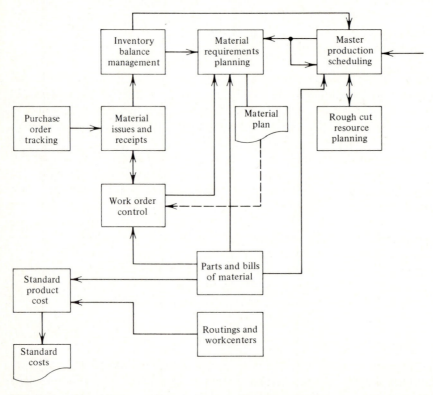

Figure 10-11 The modular structure of Materials Management/3000 software. (*Courtesy of Hewlett-Packard, Inc.*)

Examples of critical resources are labor hours, floor space, investment in work-in-process inventory, and material supplies. The RRP reports highlight the capacity constraints and help the user resolve competing demands for critical resources.

3. *Material requirements planning (MRP)* MRP simulates the complex flow of materials required to manufacture products and generates a material plan. MRP planning starts with up-to-date information about current inventory levels and the planned production requirements. Using part and bill-of-material information the material requirements for each part are calculated. The plan starts with the highest-level assemblies and proceeds through the lowest-level parts.

4. *Parts and bills of material* This module provides maintenance of engineering, accounting, and planning information about each part and product, and information on how the parts relate to one another to form the product structure (bill of material).

5. *Routings and work centers* The bill of material defines the parts and subassemblies that go into a product but does not document how the various components are assembled. The routings and work centers module maintains information that describes the locations where the products are made (work centers) and the proper sequence of manufacturing steps (routings). This information is used to generate cost information for the standard product cost module and to help develop detailed production schedules.

6. *Standard product cost (SPC)* SPC provides manufacturers with the capability of accurately calculating the standard costs associated with the manufacture of each product. The standard cost of a product is determined by accumulating all relevant material, labor, and overhead costs for the components of the product as well as the costs associated with the construction of the finished product. These standards can be used to determine product pricing and profitability.

7. *Material issues and receipts* This module helps control stockroom inventory by maintaining timely and accurate records of all actions that affect inventory balances. The data includes receipts of work orders and purchase orders, material issues from stock to a particular work order, filling of a back order, or an unplanned issue.

8. *Inventory balance management* The inventory balance management module maintains information about inventory balances and the warehouse locations where the inventory is stored.

9. *Work order control* A work order is an internal factory authorization to build a specified quantity of a particular subassembly by a specified date. All work orders require the issue of on-hand inventory for their completion. Prior reservation of on-hand inventory is the best method of preventing shortages at the time of issue. The actual issuing of parts and work orders, and the actual receipt of finished products is accomplished using the material issues and receipts module.

10. *Purchase order tracking* A purchase order represents a scheduled receipt for purchased items. Entering a purchase order requires the entry of more information than that required on a work order—for example, vendor information, shipping information, and price information.

The design philosophy of this system can be summarized as providing a functional solution that helps manufacturing management to plan and control material requirements and inventory levels. Such systems are another step toward providing full computer control for manufacturing enterprises.

10-7 THE FACTORY OF THE FUTURE

The concept of the "factory of the future" has been developed in response to the change in consumer preferences in modern society characterized by *shorter product life cycles*. The shorter cycle means more competitive products, more products being introduced, more products phasing out, and results in lower order quantities. In this sense, the age of mass production is gone and the era of flexible production is being started.

The requirement for flexible production systems dictates the specifications of the factory of the future:

Rapid introduction of new products
Quick modifications in products of similar function
Manufacturing of small quantities at competitive production costs
Consistent quality control
Ability to produce a variety of products
Ability to produce a basic product with customer-requested special modification

The core of the factory which meets these specifications is the computer-integrated manufacturing (CIM) system, which was illustrated in Fig. 10-1. The CAD/CAM process shortens the time between the concept point of a new product to its manufacturing; the FMS can produce the new product by loading a new program into its supervisory computer; the automatic assembly lines can accommodate the problem of assembling a variety of products with customer-tailored modifications; and automatic inspection maintains the high quality. All this is achieved with few workers on the shop floor. Only material-handling systems, automatic controls, and industrial robots are performing on the shop floor, under limited and remote human monitoring. Therefore, the factory of the future will not contain locker rooms, showers, and cafeteria facilities. Furthermore, automatic systems, CNC machines, and robots, the basic components of a CIM system, do not need light or heating to operate. Thus, the factory of the future will be dark and cool. Raw materials will be entering at one side, and finished products will be coming out the other end.

One of the most ambitious programs for CIM systems has been that known as methodology for unmanned manufacturing (MUM) in Japan. This was begun in 1972 and was carried on cooperatively by a consortium of Japanese government agencies, technical societies, universities, trade associations, and industrial companies. It produced a proposal and plan for developing and building, with $100 million of government funding, a prototype unmanned factory for production of machine parts and components. In this factory, a work crew of about 10 persons, instead of the usual 700, are to be employed. The proposal and plan were completed at the end of 1975 and submitted to the Ministry for International Trade and Industry (MITI) for funding. It failed to receive funding because it was considered to be too idealized a plan, and because of political pressures from two other groups, one pushing for a major program on laser machining and one for a major program on work preparation in manufacturing [10].

The result was a new program initiated in 1977 combining aspects of the laser machining and work preparation programs with that for the unmanned factory. The former MUM plan is now considered to be the idealized plan which will provide general guidelines for work in this area for about the next 20 years, while the new program is considered to be the practical plan. Three government laboratories and 20 manufacturing companies are participating in it, and this work is being guided by the Agency of Industrial Science and Technology of the Ministry of International Trade and Industry.

The program is targeted primarily at implementation of existing technology rather than having the heavy dependence on the development of new technology featured in the MUM plan. Further, one of its guiding principles is to design the hardware so that there will be a minimum need for new software. The program is still expected to culminate in a small prototype, about 750 m^2 in size, rather than a fully computer automated model factory. By 1984 (the year of termination of the project), it would employ about three or four operating workers, plus a somewhat larger support crew than that envisioned for the MUM factory. The program has been granted about $50 million funding by MITI for the 7 years. Commercialization of results will begin in the late 1980s [7, 10].

The aspiration toward the factory of the future is driven by the competitive economy in industrialized countries and supported by available computer technology. The computer control concepts that were introduced throughout this book allow manufacturing systems to become more flexible and to adapt the production process to new products in a short time. This is a substantial step toward the realization of a complete CIM system.

It seems that the efforts in developing CIM systems in the United States, Japan, and Europe will make the factory of the future more than an illusion or a dream, and it will become a reality in the near future.

BIBLIOGRAPHY

1. Allen, R.: The Microcomputer Invades the Production Line, *IEEE Spectrum,* vol. 16, pp. 53–57, January, 1979.
2. Barash, M. M.: Computer Integrated Manufacturing Systems, "Towards the Factory of the Future," ASME, vol. PED–1, pp. 37–50, November, 1980.
3. Bjorke, O.: Computer-Aided Part Manufacturing, *Comp. Ind.,* vol.1, pp. 3–9, 1979.
4. Cook, N. H.: Computer-Managed Parts Manufacture, *Sci. Am.,* vol. 232, pp. 22–29, February, 1975.
5. Federman, N. C., and R. M. Steiner: An Interactive Material Planning and Control System for Manufacturing Companies, *Hewlett-Packard J.,* vol. 32, no. 4, pp. 3–12, April, 1981.
6. Groover, M. P.: "Automation, Production Systems, and Computer-Aided Manufacturing,". Prentice-Hall, Inc., Englewood Cliffs, N. J., 1980.

7. Honda, F., et al.: Flexible Manufacturing System Complex Provided with Laser—A National R & D Program of Japan, *Information Control Problems in Manufacturing Technology*, Pergamon Press, Oxford and New York, pp. 7–11, 1979.
8. Koren, Y.: Computer-Based Machine Tool Control, *IEEE Spectrum*, vol. 14, pp. 80–84, March, 1977.
9. Kops, L.: The Factory of the Future—Technology and Management, "Towards the Factory of the Future," ASME, vol. PED-1, pp. 109–115, November, 1980.
10. Merchant, M. E.: The Factory of The Future—Technological Aspects, "Towards the Factory of the Future," ASME, vol. PED-1, pp. 71–82, November, 1980.
11. Pressman, R. S., and J. E. Williams: "Numerical Control and Computer-Aided Manufacturing," John Wiley & Sons, New York, 1977.
12. Stauffer, R. N.: Flexible Manufacturing System—Bendix Builds a Big One, *Mfg. Eng.*, vol. 87, pp. 92–93, August, 1981.
13. ———: Automating for Greater Gains in Productivity, *Mfg. Eng.*, vol. 87, pp. 58–59, November, 1981.

LAPLACE TRANSFORMATION AND TRANSFER FUNCTIONS

A-1 THE LAPLACE TRANSFORM

The *Laplace transform* is a mathematical tool that is useful in solving linear differential equations and analyzing control systems. In these fields one is concerned with transforming some functional relationship in the time domain into a new s domain, in which s rather than t is the independent variable, thus facilitating the method of computation.

A function $f(t)$ is transformed into the s domain by the transformation

$$F(s) = \int_0^\infty f(t)e^{-st}\, dt \tag{A-1}$$

Frequently Eq. (A-1) is expressed by a short-hand notation:

$$L(f(t)) = F(s) \tag{A-2}$$

For example, the Laplace transform of a constant C defined for $t \geq 0$ is calculated by substituting $f(t) = C$ into Eq. (A-1), which yields

$$L(C) = \frac{C}{s} \tag{A-3}$$

In a similar manner other useful Laplace transforms can be calculated and the results are given in Table A-1. Note that line 3 in the table is given for a special case in which all the initial conditions of a variable and its derivatives are zero.

Table A-1 Laplace transform pairs

Functions of time	Laplace transforms
1. $Kf(t)$	$KF(s)$
2. $\dfrac{df(t)}{dt}$	$sF(s) - f(0)$
3. $\dfrac{d^n f(t)}{dt^n}$	$s^n F(s)$
4. $\displaystyle\int_0^t f(t)\, dt$	$\dfrac{1}{s} F(s)$
5. 1	$\dfrac{1}{s}$
6. t	$\dfrac{1}{s^2}$
7. e^{-at}	$\dfrac{1}{s+a}$
8. $e^{-\zeta \omega_n t} \sin \omega_n \sqrt{1-\zeta^2}\, t \qquad$ for $\zeta < 1$	$\dfrac{\omega_n \sqrt{1-\zeta^2}}{s^2 + 2\zeta\omega_n s + \omega_n^2}$
9. $1 - \dfrac{e^{-\zeta \omega_n t}}{\sqrt{1-\zeta^2}} \sin(\omega_n \sqrt{1-\zeta^2}\, t + \phi)$ $\phi = \tan^{-1} \dfrac{\sqrt{1-\zeta^2}}{\zeta}$ for $\zeta < 1$	$\dfrac{\omega_n^2}{s(s^2 + 2\zeta\omega_n s + \omega_n^2)}$

Example A-1 Find the Laplace representation of the following differential equation:

$$x(t) + \tau \dot{x} = Kv(t) \qquad X(0) = 0 \qquad \text{(A-4)}$$

SOLUTION The Laplace transforms of $x(t)$ and $v(t)$ are $X(s)$ and $V(s)$, respectively. Using lines 1 and 2 of Table A-1, one obtains

$$X(s) + \tau s X(s) = KV(s) \qquad \text{(A-5)}$$

or
$$(1 + s\tau)X(s) = KV(s) \qquad \text{(A-6)}$$

A-1.1 Inverse Transform

The process of converting a function back to dependence on t is called *inverse transformation*. The inverse transform can be found by using Table A-1 in a backward direction: namely, one finds the $F(s)$ in the right-hand column and then moves to the left column to obtain the corresponding function of time. There is a problem, however: not every function $F(s)$ can be found on the right-hand column. The solution is to expand the function $F(s)$, for which an inverse transform is required, to a sum of partial functions, the transform of which is given in Table A-1. This will be illustrated by the following example.

Example A-2 The function $v(t)$ in Eq. (A-4) is a step input defined by

$$v(t) = 10 \qquad \text{for } 0 \leq t$$

$$v(t) = 0 \qquad \text{for } 0 > t$$

Determine the time function $x(t)$.

SOLUTION From lines 1 and 5 in Table A-1 the Laplace transform of $v(t)$ is

$$V(s) = \frac{10}{s} \tag{A-7}$$

Substituting of Eq. (A-7) into Eq. (A-6) and solving for $X(s)$ yields

$$X(s) = \frac{10K}{s(1 + s\tau)} \tag{A-8}$$

Using partial fraction expansion, Eq. (A-8) is rewritten as

$$X(s) = \frac{10K}{s} - \frac{10K\tau}{1 + s\tau} \tag{A-9}$$

The inverse Laplace transform is obtained by using lines 5 and 7 in Table A-1:

$$x(t) = 10K - 10Ke^{-t/\tau} \tag{A-10}$$

A-2 TRANSFER FUNCTIONS

Control systems receive inputs of one sort or another, and their response to these inputs must be determined in order to design a satisfactory system. The *transfer function* for a linear system can be defined as the ratio between the Laplace transform of the output to the transform of the input with the assumption that all initial conditions are zero.

Of special interest is the transfer function of an integral control action, which is derived from line 4 in Table A-1:

$$G(s) = \frac{1}{s} \tag{A-11}$$

Example A-3 If v in Eq. (A-4) is the input to a system and x is the output, find the transfer function of the system.

SOLUTION The transfer function is obtained from Eq. (A-6):

$$\frac{X(s)}{V(s)} = \frac{K}{1 + s\tau} \tag{A-12}$$

A-2.1 Block Diagram

One can put the transfer function into block form to perform a block diagram representation and can also recover the equation from the block. In block diagrams the output of one block is the input of the next block; such blocks are said to be in cascaded. The individual transfer functions of cascaded blocks can be multiplied to perform a single transfer function. Therefore, any number of cascaded blocks can be replaced by a single block.

A-2.2 Closed-Loop Transfer Function

In many cases the performance of an open-loop system is inadequate and a closed-loop system must be applied. In these cases the overall transfer function of the open-loop $G(s)$ is known, and the closed-loop transfer function has to be determined. The

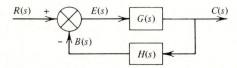

corresponding block diagram is shown in Fig. A-1. The controlled variable $C(s)$ is measured and subsequently modified at the feedback path by

$$B(s) = HC(s) \tag{A-13}$$

to have the same units as the reference $R(s)$ for comparison. The circle in Fig. A-1 is called a *summing point*, and the algebraic signs indicate how the incoming quantities are to be summed. In Fig. A-1 the summing point is interpreted to mean that

$$E(s) = R(s) - B(s) \tag{A-14}$$

Namely, the summing point is used to make the comparison between the reference variable $R(s)$ and the feedback measurement $B(s)$. The resultant quantity $E(s)$ is the error, which is used as the input to the original system $G(s)$, namely

$$C(s) = G(s)E(s) \tag{A-15}$$

Combining Eqs. (A-13) through (A-15) yields the closed-loop transfer function

$$\frac{C(s)}{R(s)} = \frac{G(s)}{1 + HG(s)} \tag{A-16}$$

Example A-4 Assume that Eq. (A-12) gives the transfer function of a dc servo-motor, where X represents the motor's speed and V the input voltage to the motor. The motor is installed in a closed loop where its speed is measured by a tachometer that emits a voltage in a rate of K_p V·s/rad. If the reference input to the closed loop is R, find the closed-loop transfer function.

SOLUTION The open-loop transfer function is

$$G(s) = \frac{K}{1 + s\tau} \tag{A-17}$$

Substituting Eq. (A-17) into Eq. (A-16) with $H = K_p$ and $C = X$, yields

$$\frac{X(s)}{R(s)} = \frac{K}{1 + s\tau + KK_p} \tag{A-18}$$

by defining

$$\alpha = \frac{1}{1 + KK_p} \tag{A-19}$$

one obtains

$$\frac{X(s)}{R(s)} = \frac{\alpha K}{1 + s\alpha\tau} \tag{A-20}$$

Since $\alpha < 1$ the closed-loop system has a smaller time constant than the open-loop one.

In many closed-loop systems, controllers are added in cascade to the open-loop system. Of special interest in manufacturing systems is the integral controller, whose transfer function is given in Eq. (A-11). A further investigation could show that for a constant reference input R, the integral control action eliminates any steady-state error in the closed-loop system.

BIBLIOGRAPHY

1 Harrison, H. L., and J. G. Bollinger: "Introduction to Automatic Controls," 2d ed., Harper & Row Publishers, New York, 1969.
2 Raven, F. H.: "Automatic Control Engineering," 2d ed., McGraw-Hill Book Company, New York, 1975.

INDEX